SURVIVAL OF THE FITTEST

Mike Stroud completed a degree in anthropology and genetics and qualified as a doctor in 1979. He became a Member of the Royal College of Physicians in 1984 and a Fellow in 1995. He has travelled widely – in the Himalayan regions of Ladakh and Zanskar and in North Africa and South America. He was the doctor on the 1984–86 'In the Footsteps of Scott' Antarctic expedition and has been a medical officer with the British Antarctic Survey. He first teamed up with Ranulph Fiennes in 1986 for the first of their three attempts to journey on foot to the North Pole unaided. He went south again to Antarctica with Fiennes in 1992 and in 1993 was awarded the O.B.E. in recognition of their first unsupported crossing of the Antarctic continent from coast to coast. After six years in the 1990s as Senior Research Medical Officer advising the Ministry of Defence on nutrition, exercise performance and survival, Dr Stroud returned to hospital medical practice with a particular interest in nutrition and metabolism. Nevertheless, he has continued to undertake a variety of gruelling challenges, including Eco-Challenge multi-sport adventure races. Recently, he and Ranulph Fiennes completed seven marathons in seven days on seven continents. He has contributed articles to many international scientific publications.

Mike Stroud

SURVIVAL
OF THE FITTEST

Understanding Health and
Peak Physical Performance

YELLOW JERSEY PRESS
LONDON

Published by Yellow Jersey Press 2004

4 6 8 10 9 7 5 3

Copyright © Mike Stroud 1998, 2004

First published in this edition in Great Britain in 2004 by Yellow Jersey Press
First published in Great Britain in 1998 by Jonathan Cape
Random House, 20 Vauxhall Bridge Road,
London SW1V 2SA

Random House Australia (Pty) Limited
20 Alfred Street, Milsons Point, Sydney,
New South Wales 2061, Australia

Random House New Zealand Limited
18 Poland Road, Glenfield,
Auckland 10, New Zealand

Random House (Pty) Limited
Endulini, 5A Jubilee Road, Parktown 2193, South Africa

The Random House Group Limited Reg. No. 954009
www.randomhouse.co.uk

A CIP catalogue record for this book
is available from the British Library

ISBN 0-224-07507-1

Printed and bound in Great Britain by
Bookmarque Ltd, Croydon, Surrey

Contents

	Preface	ix
	Introduction: The Dawn of Fitness	1
1	Facing the Challenge	5
2	Out of Africa	29
3	A Gene for the Fastest	36
4	The Happiness of the Long Distance Runner	50
5	Crossing Antarctica	66
6	Marathon of the Sands	88
7	The Heart of Darkness	122
8	Cries of the Heart	151
9	Survival of the Fattest	163
10	Fit for Life	178
	Conclusion – In Sickness or In Health	203
	Postscript – Running the World	
	Seven Marathons, Seven Days, Seven Continents	209
	Sources and Further Reading	241
	Acknowledgements	245

Illustrations

1 Helen Klein, aged 72, enters the first Eco-Challenge race in Utah (by Mike Stroud)
2 Moving through flooded canyons (by Robert Houser)
3 Linford Christie (by Mike King)
4 Liz McColgan (by Mike King)
5 The London Marathon (by Mike King)
6 Sir Ranulph Fiennes in Antarctica (by Mike Stroud)
7 Mike Stroud caught in a crevasse (by Sir Ranulph Fiennes)
8 Marathon of the Sands (by Mike Lean)
9 Medical tent in the Sahara (by Mike Lean)
10 Working in Antarctica (by Mike Stroud)
11 Mike Stroud with Emperor penguins (by Roger Mear)
12 Chimpanzee, man's closest genetic relative (by F. and K. Ammann, Planet Earth Pictures)
13 Obesity (by Dan Goodrich, Picture Group, Colorific!)
14 Canadian Eco-Challenge team – Sir Ranulph Fiennes, David Smith, Rebecca Stephens, Mike Stroud and his father, aged 70 (by Moyra Wiseman)
15 River crossing by rope (by Kevin Gilbert, Corbis)
16 Mike and Vic Stroud after the race (by Rebecca Stephens)
17 South American marathon (by Gill Allen)
18 The New York Marathon (by Gill Allen)

To my Father
who introduced me to the great outdoors.

Preface

THIS BOOK is about the advantages and disadvantages of our evolutionary inheritance. On the positive side, we all have a resilience and physical strength that far exceed people's expectation. On the negative, there is a propensity to develop serious health problems.

There are many ways in which to push the human body. Some of the chapters relate to performance in the challenges of sport – the activities that men and women have evolved to take the place of hunting and gathering – while others concern our abilities to cope with extreme conditions, examining them through tales of those who have sought challenges in the wilder parts of our planet, or who have met life-threatening situations through accident or misadventure. In addition to descriptions of human strength, some chapters cover human weakness – the heart attacks, strokes, diabetes, and other illnesses that arise from our modern lifestyles and diets that are far from those for which evolution designed us.

The ideas expressed in these pages were developed through an education in anthropology and medicine, and a career as both a hospital doctor and researcher into human performance and survival for the Ministry of Defence, but they are illustrated by more unusual personal experience. I have been fortunate enough to take part in many physically demanding events and expeditions in some of the most savage and beautiful regions of the world. These gave me the chance to explore my own and others' physical and psychological limits, and have led me to the belief that by taxing some of our hidden evolutionary strengths we can avoid many of our inherited weaknesses, and so remain active and healthy well into old age.

Mike Stroud
January 1998

*

The Dawn of Fitness

IT WAS a bitter morning. The sun rose with some promise of radiance, orange rays glinting on the hoar frost, but a few minutes later the group entered the marshlands where wreaths of mist emanated from dark unfrozen water. The herald of warmth became a mere dull disc hanging in the misty silence, and the deep cold persisted. As the morning passed the sunlight became no stronger, although its pale presence made the task of tracking easier. The hunters moved forward at a greater pace, running mixed with walking, depending on terrain. The running was welcome – at least it kept the chill at bay.

The men had been following the beast for the best part of three days, moving as fast as they could over the bare, snow-covered plains or pushing through huge areas of naked birch woods. Now they were chilled, tired and hungry – desperately in need of food to replace the fuel that their muscles had burned away. They had seen little that they could eat since starting out. A few nuts, one or two fungi, that was the sum of it – not enough to sustain them if the hunt were to last for many days more. Nevertheless, they still felt reasonably strong, drawing on their bodies' fat stores that had been laid down months before when the hot summer sun had shone and the land had been a place of warmth and plenty. All that now seemed far away – distant dreams of comfort to offset pain as they waded across yet another icy river.

In their dreams, the whole of summer was a time of pleasure but, in reality, it had been far from easy. Well before the sun had reached its annual zenith, good water had become hard to find and their long trips in search of food had been very difficult. Two of the tribe had succumbed to the heat even before the day of shortest shadows, and more had died later after the return of longer nights. Yet, as with all painful memories, the truth of that time was now relegated to the subconscious. Their thoughts dwelt now on warmth, comfort and recollections of a bounty of food.

Grubs had been plentiful that summer and nuts and berries had coated the trees right up to the first frosts. As well as getting stores into their huts for winter, everyone had been able to eat well right through the leaf-fall, putting on extra fat which now gave them a layer of insulation against the seeping chill and provided a reserve of fuel to keep them moving. Without it, the hunters would have been forced to give up days before.

The mammoth had been wounded from the very beginning, caught by their spears as it wandered into a carefully prepared trap. But although maybe four or five of the sharp flint blades had found their mark, they had served only to weaken the animal. Now, nearly four days later, it still showed tremendous endurance and there were few signs of it slowing down. That was commonly the way. Every winter, when the persisting snows finally forced the men to try to hunt for food, they would go out in their groups and often return empty-handed. They almost never killed anything cleanly, and when they did try to bring down a mammoth, usually they merely injured it and then spent many days harrying the beast before there was any chance of success. Often, they would lose their quarry in the marshes or become so exhausted from the chase that they had to give up. A mammoth was a huge prize but one that was never gained without great cost.

On this occasion, the hunters had not seen the animal since the previous evening and all they could do was to follow the trail that wound through the soft snow. The men were flagging, but they could tell from both the sharpness of the tracks and the recently bent reeds that the beast must also be tiring. They were definitely gaining on it. With this realisation came the onset of fear as each began to think of the final confrontation. Past experience did not make it a prospect to relish, but the burst of adrenaline that fear induced within them renewed their strength, and the line of men moved across the land in a steady lope. Suddenly, there were excited calls from the left. Two of the group were shouting, pointing with their spears and waving their clubs. They had obviously seen the animal not far away. The prize was close – food, fur and bone were there for the taking, and their families would be thrilled if they returned well laden. It would take many days to get back with their loads but at least all their efforts would have been worthwhile.

A few hundred yards away an area of high reeds was active in the cold and windless afternoon. The men on the hill were pointing there, clearly able to see their quarry, and those down below paused to assess the situation. The oldest member of the group had been to this place before and quickly described it to his companions. He knew that in front of them the marshy ground became a thin peninsula, a finger of ever softer mud that extended out into a deep dark lake. It was a strange mere, a pool with a vile taste that he did not understand and one which for some reason remained unfrozèn through most winters. The mammoth would be trapped by open water on three sides. It was an ideal place for victory.

The men from the hill rejoined them and they moved off in a line, edging warily along the narrowing mud banks, their stone-tipped spears ready for the kill. They had no misconceptions about the effectiveness of their weapons and they needed space to manoeuvre if the animal turned on them. As they approached the narrowest point of the isthmus, they were particularly vulnerable and so were keen to cross to where they could see that the banks widened out into an area filled with reeds. At the same time, they needed to move slowly in order not to panic the beast. On the wider head of the peninsula they would have their best chance. It was now crushingly silent and each man wondered what the animal would do next. Their hearts beat palpably in their chests and their bodies became as taut as springs.

Without warning, and before they reached the wider reed beds, the mammoth re-appeared. Trapped, wounded and enraged, it charged, determined to rid itself of the persistent pests that had stung and chased it for so long. Although they were prepared, the huge animal charged so fast that there was little they could do to avoid it. Confined by the mud banks, they were trapped as surely as they had meant to trap their prey, and they knew that now no number of spears would stop the mammoth in its tracks. The realisation that they would be crushed brought immediate action: they all turned and fled. It was nearly two hundred spans back to the wider lake shore where rocks and trees would offer some protection – much too far. While only halfway to safety, their legs began to fail them and the pounding of tree-thick limbs drew close behind.

Most of the hunters made it, throwing themselves from the mammoth's path just as they thought they could run no more. Two did not. One, a young man of only sixteen summers, was felled as he tried vainly to keep up. He had never been one of the faster runners and some had said that he should never join a hunt. Now they were proved right as, his head crushed like fruit, he met his end, but at least it was quick and merciful. The other victim, the older man who had known the place before, was not so fortunate. At fifty or sixty summers, he was slower now than when he had been in his prime and perhaps he too should have stayed at the camp, although that would never have been practical. It was many years since the tribe had last travelled in the region and he was the only one of the elders who remembered it well and could still move fast enough to go out with the hunt. At least, nearly fast enough, for although he had not been killed outright, he was far worse off. His chest had been crushed by one of the mammoth's feet and he lay on the ground, panting for breath and gasping with pain, while the great beast disappeared in the gathering twilight.

The uninjured men emerged from the rocks and examined their two companions. The younger, quite dead, could be left but the older man was more of a problem. There was nothing they could do to help him, and if he did not die from his injuries he would die from the cold. Nevertheless, they could not abandon him to the wolves and felt an obligation to try and get him back to his family. A stretcher of branches and reeds was hastily constructed so that they could drag him slowly home. To make their journey would take all of seven sundowns, perhaps more, and unless they were lucky there would be little or nothing to eat on the way. It would consume all their strength and determination, but each man knew that they had no choice. Each had survived much worse before.

ONE

*

Facing the Challenge

For many people, the marathon represents the ultimate in human endurance, but perceptions are changing. Not long ago, the ability to complete the distance was thought to be a capacity confined to superhumans alone, but now it is accepted that ordinary people can complete such races. It is a fact witnessed by the tens of thousands who run through the streets of our major cities every year. Indeed, many now see the marathon as at the lower end of the endurance spectrum. Where, for decades, there were only a few scattered races providing greater tests – the Tour de France or the Hawaiian 'Iron Man' – numerous ultra-distance events are now staged all over the world. Simultaneously, there has been an increased interest in activities such as mountaineering and prolonged expeditions to far flung regions. From city streets to mountain tops, more and more people are prepared to pit their fitness and mental stamina against prolonged physical demands. As they do so, it becomes clearer that both men and women have extraordinary endurance capacities.

I first met Mary Gadams in 1994 while running in a multi-marathon across more than 120 miles of Saharan sand. After that race, she read my book, *Shadows on the Wasteland*, which described my crossing of Antarctica with Sir Ranulph Fiennes. That was a journey made on foot, without re-supply of food or fuel, and with no help from other men, animals or machines. At the time, it was the longest, self-sustained walk in history. Taking this as a proof of my doggedness, Mary became convinced that I should be one of her team as she hatched plans to enter the first ever Eco-Challenge. She telephoned from the United States one quiet Sunday afternoon – domestic routine and the indulgence of a late roast lunch shattered by her invitation to wake up, train, and push myself hard once more.

The Eco-Challenge was a race conceived to take competitors beyond their limits. It would cross more than three hundred miles of demanding back-country in the south-west of the United States using a variety of modes. Much would be running, hiking or mountain biking, for which I could offer both navigational skills and a reasonable level of endurance fitness, but the race was also to include horse-riding, Canadian-style canoeing and white-water rafting, at all of which I was a complete novice. Even the rock climbing, at which I counted myself quite adept, would include the use of artificial techniques to climb huge rock walls. Ascending clamps, pitons, and tape steps, all were unfamiliar to me. I was intrigued but daunted.

Each team had to comprise five persons, with at least one male and one female, and all participants had to complete every event. Here was another cause for concern for, despite my proven slogging skills, I did not see myself as in the same athletic league as her other choice of team mates. I had seen Mary herself running in the Sahara and knew that she was better at great distance than I was. Bill, who had already agreed to join her, had come second in the U.S. National Triathlon Championships, which made him pretty exceptional, and Whit, the other male, held the record for the fastest ascents of the highest peaks in every U.S. state. Most extraordinary of all, was the second woman in the team. Helen Klein was one of the greatest distance athletes the world has ever known.

When Mary persuaded Helen to join her, Helen was already 72 years old and a great-grandmother. Nevertheless, she was as fit as, if not fitter than, most people fifty years younger. She had already shown a phenomenal prowess at extreme distance running – a sport that she took up at the ripe age of 55 to compete with her distance running husband who was fourteen years her junior. I guess that that age discrepancy in itself denotes an unusual biological phenomenon. Helen was always young for her years, and as she became older, the difference between the calendar and reality went on increasing.

After taking up jogging, Helen steadily built up to running marathons. Soon she could beat all comers of her age, and so she progressively increased her distances to discover that she could also perform well at much longer events. Here she could provide competition to anybody, whatever their age or sex. In a way, this was not surprising, for it has been known for many years that beyond the non-stop 100-mile mark, women are as good as men. However, it is unusual to find top-class ultra-distance competitors over the age of 50 and unheard of for an athlete to continue successfully into their seventies. Helen broke all rules. By the time she joined Mary's team for the Eco-Challenge, 17 years of running had led her to completing more than 75 marathons and 150 ultra-marathons.

Even more worrying for me was that Helen showed little sign of slowing down with her advancing years. She could still run 100 miles in a little more than twenty hours – the equivalent of running close to four consecutive marathons at a pace that was respectable for a young woman. As a veteran, she was truly in a league of her own. Her personal motto was 'I'd rather wear out than rust out', and although I was only 40, I would clearly have trouble keeping up with her. Yet, despite my expressing such reservations, Mary was persuasive. I agreed to go for it and, following a now very late lunch, went out for a lengthy jog. The experience cost me violent indigestion and several days of stiffness. There was a long way to go but life felt good.

*

The horses were brought down from the hills on the evening before the race began. You could tell from their wild neighing and the sharp reports as they kicked at the fencing that it had been a long time since they were last cooped up. Speaking to the wranglers, we learned that during the whole winter and the spring the horses had run wild on the high pastures. It did not bode well for a docile mount, and when I walked round the corrals to see for myself that these were the 'fully trained trekking horses' described by the race organisers, I was far from reassured. A Palomino in particular caught my eye. Bucking, kicking and even biting at its companions, it looked as if it had never been broken-in, let alone been used for hacks. It also possessed the strangest eyes I had ever seen on a horse – blue, almond-shaped and unearthly pale against the browns of its nose. They looked as if they came from a wolf rather than any equine ancestor, and I prayed that our team would not be making closer acquaintance.

My prayers were not answered. As dawn broke the next morning it was a strange scene that filled the slopes of the desert-brown hills. The fifty teams of five persons each were now mixed with the one hundred and fifty horses that were even more excited than on the previous evening as they were saddled and assigned to riders. The first stage of our race was to be about twenty-five miles with three members of each team riding while the other two ran alongside. Our route stretched away from the high ground on which we stood across mainly flat scrub-covered country dotted with 'buttes' of red, towering sandstone rock that looked liked movie-set castles. Those familiar with horse-riding well knew that the stage would be hard on man and beast while the uninitiated, like myself, still thought it would be easy if rather frightening.

Although in the few months before joining the team I had considerably improved my equestrian skills, this was no great feat in view of my starting point. My only previous experience with horses had been a brief attempt to mount one many years before, and that had ended with my falling straight over the far side as the animal shrugged me off. It had seen a novice coming and decided that it had no interest in even walking with such uncoordinated baggage. Now, after attending evening riding classes, I believed that I could get on without such an immediate exit – until I saw that my mount was to be the horse with the ghostly eyes. At that moment I backed down and suggested a change of plan. It seemed safer for one of the more experienced equestrians in the team to take charge of this wilful beast, and since neither of the other two horses looked much better, I was to start out running. In doing so I would join Helen, who was also an inexperienced rider.

We got away at 6 a.m. – not an orderly start but a wild stampede. In retrospect, this was inevitable. The horses were unused to being saddled and mounted and many of the riders were of a poor standard. Why nobody was seriously hurt I do not know. As the countdown was completed, one hundred and fifty animals took off across the desert in a frenzied galloping group, most of the riders totally out of control and hanging on for dear life. With the rough going it was only moments before there was a scattering of bad falls. Some of the contestants dropped at the feet of horses coming on behind; others became entangled in the improvised straps and ropes that were used to attach their kit and were dragged along as bouncing sacks. It was surprising that nobody was seriously hurt, and after about ten minutes, the horses became calmer and spread out as things settled down. The teams which had not sustained a fall regrouped and went on, the rules insisting that all five members stayed close together at all times through the ten days that lay ahead. Behind us, small groups were dotted over the starting plain, some nursing their bruised team mates, others waiting as horses and their supposed riders circled each other warily in a taut-reined waltz of mutual suspicion.

Once the horses had settled, both my own and Helen's courage returned and after about ten miles we swopped with Bill and Whit and took to the saddle. The horses were still restless, and I could not get my mount to trot for any distance without him getting progressively faster. Before long I would find myself at a canter or full gallop, which terrified me. Then I realised that I had all the space in the world and it did not seem to matter. Careering across the open desert was a very different thing to cantering along some confined British bridle path and, with American-style saddles, it was more comfortable to go fast anyway. Rising to the trot was virtually impossible with an armchair for a seat and so slow speeds, other than a walk, gave a very uncomfortable ride.

Unfortunately, the horse refused to stand still at any time. This was a problem because the team had to stay close together, but every time I tried to wait for the others to catch up, my horse would start pawing the ground, snorting and threatening to rear right up and rid itself of the pest that kept holding it back via the bit. Within a few seconds of stopping, I would be forced to kick the animal on to stop it throwing me. In the end, the only way I could keep in touch with my companions was to gallop away from them for a couple of minutes, and then back towards them again. I had to do this all the way to the first checkpoint at the 14-mile mark.

At fourteen miles we were halfway through the first riding/running stage and there was a veterinary inspection. The vet was unhappy about the way that my horse was overheating, probably due to the considerable extra distance we had travelled together. To be safe, he pulled it out of the race, and so we set off again with just two animals. This was actually no disadvantage. Many of the teams who did not lose horses that were weakening were to regret it later. At the pre-race briefing the night before, it had been pointed out that any team was only as fast as its slowest member, but few had recognised that this was likely to be one of the animals. Although men and women are clearly much slower than horses over a few miles, when it comes to tens of miles, things can change. Humans make incomparable endurance machines.

After the checkpoint, Whit, Mary and I took to running and the terrain became more tiring. We were gaining height up a gently winding valley, entering a range of barren, orange-cliffed hills. It was a glorious place in which to be. Extraordinary natural rock walls and pinnacles were made mysterious by evidence of past Indian settlements. On one cliff beside the trail aboriginal art figures were carved in the ochre stone. You could feel the eyes of history gaze down as we passed beneath.

It was a little way past these carvings that we encountered our first real difficulty. Helen wanted to stop her horse and get off for a pee. She asked me to hold the animal while she retired briefly behind a rock. The horse, that ghost-eyed Palomino once more, was docile enough until she came back to reclaim it. I was holding it by the bridle as she started to climb back into the saddle, but as she did so, all hell broke loose. The horse seemed to think that its day's work was over, for as soon as Helen

9

got a foot in the stirrup it reared, neighing, whinnying, kicking and bucking wildly. Helen was thrown some distance but landed unhurt. I, on the other hand, felt my head meet one of the beast's flailing hooves and immediately blood trickled down the side of my face. I wanted to let go and try to staunch the flow, but the horse set off across the valley floor, and I knew that it would go for ever if I did not hang on. Losing your horse meant immediate disqualification, so I kept hold as it dragged me forward, thrashing its head from side to side to try to rid itself of my unwelcome weight.

The whole episode lasted for only twenty seconds or so, though it felt much longer. Holding a galloping horse while being dragged across stones and through scrub turned out to be terrifying but, once committed, letting go seemed an even more frightening option. At least while I held the bridle I could keep my head clear of the pounding hoofs and rapidly passing rocks, and even the Palomino eventually realised that dragging a man by the mouth and halter was not much fun. It finally came to a halt, maliciously lining up with a large and vicious thorny bush with which I now made intimate acquaintance.

Whit ran up and took the halter while I struggled to free myself from the spines. He looked startled when he saw the vivid scarlet of my previously white head-band, with the blood running down from beneath it to my jaw. The blood was already beginning to congeal and must have looked as if it covered a deep wound down half my face. Whit seemed reassured when I spoke and wiped some of the blood away. Although the cut was just a scalp wound an inch or two in length, it was deep enough to need some stitches.

They would have to wait. The rules allowed medical help at the camps between different stages but not any assistance out on the course. We had to cover the rest of the distance quickly so that I could get my cut cleaned and fixed up. With a headache and bruises, and feeling as if I had been through a couple of rounds with Mike Tyson, I went back to the trail – running, loping, walking and finally riding once more to make the first full staging point in the early afternoon. There I handed over my horse to our two-person support team and received seven stitches in my scalp, inserted with only perfunctory local anaesthetic. It was uncomfortable but had the advantage of speed. Night was just a few hours away and we had to move on. We ate and drank well and then took on what we thought would be adequate supplies to see us through the second stage. It was to be a long run-come-hike through deep canyons in the range of mountains we could see ahead before once again crossing open, scrubby desert.

We set off as soon as we could so as to put as much distance as possible behind us before nightfall. Part of the canyon route was a 'dark zone', considered to be too dangerous to negotiate without light. We wanted to reach the start of it before resting so that we could enter the difficulties first thing the following morning and then complete the canyon section by the evening. In that way we would be able to cross the hot desert in cool darkness and, if we went without sleep, make it to the end of Stage 2 by early the day after. With the entire distance looking like less than one hundred miles on the map, we reckoned that under 48 hours was a realistic time. As it turned out we were hopelessly wrong, although all the other teams seemed to make the same mistake.

Soon after leaving the checkpoint, we entered the canyon system. We were still a little way from the difficulties that would be barred in darkness, but even here the canyons were winding, narrow, dark and, worse, were largely filled with bitterly cold water. The water was deep enough to force us to swim for quite long distances, pushing our backpacks with our food and emergency equipment ahead of us, shivering violently as we did so. Furthermore, where the canyon bottom was not flooded, the way was jammed with rock falls covered in slimy mud. It was not, therefore, until after midnight that we arrived at the control point at the start of the difficult section and lay down to rest on a narrow ledge above the river, and it was only two or three hours before we were up again to eat and pack before daylight.

It took the next day to struggle through the dark zone gorges. They were so narrow that it felt as if we were in caves rather than clefts. We swam much of the way, and by the time we had completed the difficulties we were utterly exhausted, and there were still many more miles of easier canyons to go. In the end, a second night was nearly over before we emerged mud-coated and somewhat hypothermic at the desert's edge. We were only halfway to the second stage-point, and later in the day we would run out of food if we did not cut back. We decided to spin it out, and with no breakfast set off once more. The mountains were soon left behind and the bad-lands surrounded us. Above, the sun rose into a disturbingly cloudless sky.

The desert was another nightmare. There were a host of river springs and water holes shown on the map but it turned out that they were either mythical or poisonous – marked with those skulls on posts seen in old Western films. As we set out from the next control point a race official warned us that only Well Springs was safe to drink from. In any case, most of the others were dry. With these springs more than forty miles away and the temperature rising to 30°C in the shade – with no shade anyway – we could not carry enough water to run the distance. We had to walk rather than trot, and it was going to take a very long time.

The terrain was also more difficult than we had anticipated – little easier than the canyons left behind us. Our maps suggested that the desert section would be large open plains, interspersed with empty smooth-floored mountain passes and some truly giant but dry canyons. The contour lines on the maps, however, did not reveal that the entire route was criss-crossed by narrow, steep-sided dry gullies which constantly sent us in search of a safe way down into them and then back up often vertical loose far sides. It made progress very slow indeed. Having risen at 6 a.m. from just one hour's sleep by the side of the last river, we had not reached the springs by 2 a.m. the following morning, and were still less than halfway to the next checkpoint and more food supplies.

Late that second night we ran out of water as well as nourishment. We were parched, and with the springs now less than ten miles away, we all wanted to press on. Helen, however, was in no state to persevere. Although her endurance was living up to her legend, she was used to both eating and drinking on the run and resting between long days of effort, and the lack of food and rest were beginning to tell on her in a way she had never experienced. Seeing her sway down the trail in the darkness, we had no choice but to halt. But we had no grounds for complaint. We were still in the top twenty per cent of the field, and when we met other exhausted teams, the sight of our great-grandmother made their jaws drop. We loved her for it.

*

There is a common belief that prior to this century everybody died young because the state of medical knowledge was so poor. It is a misconception. If our ancestors reached adulthood, there was a fair chance that they would reach the middle-age of our standards and some would even survive into old age. After all, three score years and ten is a term from the Bible. That is not to say there has been no increase in the *average* human lifespan. Clearly there has, although most of the increase followed changes in public wealth and health rather than any improvement in medical practice. The introduction of clean water and universal sewerage in the Victorian era did more for average longevity than anything doctors have ever produced. The increase in average age of death also reflects a decline in the mortality of infants, children and women in childbirth rather than of older individuals. Throughout history, and even pre-history, the life expectancy of a forty-year-old has not changed as much as most people imagine. At least, it did not do so until the last twenty years or so.

In these last two decades, the situation has finally altered. Recent medical advance has now influenced the lifespans of the elderly enough to make a difference. In nations with high health spending, the three score years and ten have moved up to and beyond a round four score. But why has this change come so late? The modern medical era should have borne useful fruit well before the last twenty years, particularly with the discovery of antibiotics and their successful treatment of most infections. It seems that every useful medical discovery was offset by detrimental changes to health occurring with alterations in lifestyle. Although we learned to cure diseases that had once been fatal, new and deadly conditions moved in to replace them. Instead of dying from lungs infected with pneumonia or ravaged by TB, we succumbed to bronchitis and lung cancer related to our smoking; instead of cholera and typhoid in our gut, the incidence of bowel cancer rose; and instead of rheumatic fever damaging our hearts in childhood and then killing us later through the scarring of our heart valves, we perished from heart attacks as our arteries became clogged. Overall, millions of us continued to meet an early grave – not through the natural hazards of life, but through smoking, the wrong diet and a lack of exercise.

★

We lay on the desert track with the sky above veiled with stars, although their beauty was hard to appreciate when legs and bodies were so very tired, uncomfortable and sore. We carried sleeping bags for the cold nights but the ground was hard and stony and our tongues cleaved to arid throats, coated in sandy dust. As we tried to rest in the darkness, we could comfort ourselves with the knowledge that the springs were now quite close. We would stop for just two hours for Helen to recover slightly and then move on well before sunrise. In the cool of the early morning, we would be able to lope along, and a couple of hours at most would see dawn at the springs.

Well Springs – what visions that name engendered. Streams tumbling down grassy hillsides, water bubbling over stones and pools in which we could bathe our painful feet. Beside them we would sit and share the very last of our food – a one person pack of breakfast cereal. Then, fully refreshed, we would push on over the last part of the desert to the next staging point. With luck we would be there by that evening, and then we could eat properly.

Yet again, hopes were not matched by reality. Although we arrived at the site of the springs quite quickly, we could find no water. At first we thought that we must have been mistaken about our position and spent some time taking bearings from dimly-seen hills. These confirmed our grid reference. We stood and listened carefully but could hear no running water, just the crushing silence of isolation. Quartering the area carefully, we came across another team that had arrived the evening before. They too had failed to find the water and were waiting for morning light. Ominously, they had already unpacked their emergency radio to call for help if they could not find water. One of them was in a bad way from dehydration, and they knew they could go no further.

The sun rose quickly, and once there was more light we began the search again. After half an hour or so, we heard a shout from some distance. Down a small side valley someone called 'It's here!' We all ran towards the cry – to see, to drink, and to relax. What disappointment. Instead of even one rippling brook, one burbling waterfall, or one small pool, there was just a puddle – a puddle a couple of yards across and only a few inches deep. To make matters worse, it had been trampled by cattle and the water was filled with their droppings. It was a ghastly addition to a liquid that was already seething with small larvae and other swimming creatures. A hard decision had to be made. Should we give up here or extend our trust in our sterilising tablets a long way beyond their specification?

It crossed our minds that we should stop – but that was not the answer for which we had come so far. Not only had we suffered the days of racing already achieved, but we had been through all those months of training and preparation. We could not simply give up because of squeamishness. In silence we filled our cups, dissolved our tablets and tentatively sipped the resulting vile, brown, iodine-flavoured filth. Then, still silent, we prepared our bottles for the rest of the day and made ready to go on. Only Helen actually put the stuff on her small share of muesli and ate it while chatting amiably to the other team who sat in abject depression. Only our 72-year-old faced the situation with unflagging humour and then brightly called for the off. As we left, the others called HQ to throw in their towel.

Our second day across the desert went no better. We walked through the heat, drinking reluctantly from our still limited water supply and heading for the next checkpoint where there would be bottles of mineral water and cans of coke to refresh us. But again, the maps and terrain matched poorly and so a nightmarish zig-zag took the remaining twenty-five miles up nearer to fifty. We had to rest again, and lay that night by the trail in a dusty canyon. It was lunchtime the next day before we eventually reached the second staging point. Instead of 48 hours from the first, it had taken four days and nights, and by the time we got there we all had bad stomachs and diarrhoea from the contaminated water. Even so, we were still in the race, while several others had dropped out. Many of the teams had been unable to face the cowshit water.

We remained at the second stage-point for a couple of hours while Helen was patched up enough to go on. Although young in heart and limb from years of exercise, she was not so young in skin. Hers had seen a long life of Californian sunshine and so her skin was fragile and paper thin. Now it needed to be bandaged to cover the cuts, blisters and cacti scratches carried with her.

The next stage, on mountain bikes, covered about thirty miles of unmade tracks to reach the edge of another canyon system. The ride went well for the first couple of hours, but then disaster struck. Heading down a steep winding hill on a loose gravelly surface, Helen lost control – a result of her gamely performing another sport with which she was hardly familiar. She hit the ground just ahead of me and I braked to a stop. She was quite still and did not stir for several seconds. I was off my bike and at her side before her eyes even flickered.

As a doctor, I knew what I should do. This was a woman who had hit her head hard enough to lose consciousness, albeit briefly. Such a patient should be admitted to hospital for observation, for a blow to the head can cause serious problems later. If there is slow bleeding within the skull, the pressure can build up without signs until quite suddenly the victim can lapse back into unconsciousness. There may then be only minutes before the pressure rises enough literally to squeeze the hind part of the brain out through the hole in the base of the skull. If this happens, it is fatal. Helen had been out for a few seconds, but afterwards was entirely lucid. She said she felt fine and had no headache. While she was raring to go, I faced a dilemma. I knew what I should do but, as a contestant, I did not want to do it. In the end I left it up to Helen herself. She had been a nurse in younger days and knew exactly why people went to hospital in such circumstances. It did not surprise me that she opted to go on, asking me to keep an eye on her through the night to ensure that she did not lapse into a coma. I agreed, and as soon as I had performed a little first aid we were once again on our way.

Above one eye and down one leg, Helen displayed some impressive lacerations, pouring blood. They should have been stitched but that was not an option. I cleaned the wounds as best I could and applied paper adhesive closures carefully. They would leave her with scars but she would bear them with pride. The whole episode served to emphasise her extraordinary mental as well as physical strength.

★

In addition to the misconception that in the past people died young, another false premise directs our thinking to the likely consequences of ageing. In western societies, it is accepted that as decades pass we inevitably become less active. It is assumed to be a natural part of senescence but, from an evolutionary point of view, this cannot be true.

The events I described in the prologue – the multi-day chase of the mammoth – were pure fiction. They were guesswork, portraying one aspect of life from perhaps ten thousand years ago. The image fits the popular pictures of Stone-age life in northern Europe, but anthropologists believe that the idea of early men as a spear-wielding mammoth hunter must be misleading. Women probably hunted also and, away from winter, the hunt would have been a minor part of life anyway. Antelope are very fast and wary and, as Jared Diamond has pointed out in his book *The Third Chimpanzee*, poking a buffalo, rhinoceros or mammoth with a stone-tipped spear would be tantamount to suicide.

Most of the early hunter-gatherer populations must have depended more on gathering than hunting. Grubs, worms and insects would have been added to a diet of fruits, nuts, roots and berries, and so they were not vegetarians. Rodents and birds would also have been caught on occasions, but success with larger quarry must have been rare. The return of the hunting party bearing rats and rabbits is a somewhat less macho image than one bearing elk and elephant.

Nevertheless, gathering foods would have entailed hard work. Men, women and children must have travelled widely and almost constantly, searching for plants, tracking with the hunts that did occur, and crossing whole regions with the change of seasons. Activity, extremes of climate and an almost constant threat of starvation must have been an integral part of life. To survive, everyone must have been extremely physically fit, whatever their age.

The need to stay fit would have begun to change with the advent of civilisation and the development of farming. Early in the process, most people would have remained physically active, working on the land. Later, however, more and more individuals would have become more sedentary as they took to trading and administration, and not until very much later, with the arrival of industrialisation and transport, would levels of activity have markedly diminished. Only in recent years, in the age of silicon, has inertia become the norm. A century ago, most people would have been active through most of their childhood, teenage years and adulthood, but in the last few decades the situation has been transformed.

Today activity levels for most individuals diminish with age in line with their changing commitments. While young children, our games keep us active for much of every day – although even this is threatened by the lure of the 'Megadrive'. Then, as we go through high school, spontaneous games decline and some days see little in the way of vigorous movement. Early adult jobs or university education see activity become more sporadic still, confined to an hour or two a week at a favourite sport. Then it stops altogether. In the developed world, most men and women over the age of 30 abandon physical exertion altogether and turn instead to the inescapable attractions of the motor car, television and the Internet. At around the age of forty, some do briefly resume exercise in an attempt to maintain some semblance of a youthful body, but few stick at it for long. Soon they rejoin the sad but inexorable drift towards sloth – the normal four-stage progression of accepted adult ageing described by Sue Lim in a BBC radio programme as: 'Lager, Aga, Saga and Gaga'.

When I make this point to friends or colleagues, I am accused of exaggeration and, yes, I admit, many adults do undertake something. They tend a garden, perform some D.I.Y., push trolleys round the supermarket, and may even walk the dog. Yet none of these things will stress the heart, lungs and muscles in ways for which they were designed. Even if spared immobility through bad joints or other illness, middle to old age sees those who do not have physically demanding jobs lose their physical fitness to such a degree that they even fail to meet the minimal demands of life in the late twentieth century. They become too infirm to get by without help. Of course, they may say 'So what? We have no need to push ourselves at all.' Few realise that they are turning a blind eye to the advantages that exercise can bring. Meanwhile, stories of capacities such as Helen's become more and more extraordinary.

*

The end of the cycling delivered us to the lip of Snake Canyon – another dark zone. In many ways we were grateful for that and settled down for our first night's sleep of more than a couple of hours.

Unfortunately, it soon began to rain and the bivvy covers of our sleeping bags – being thin to minimise the weight of our backpacks – were totally inadequate for a desert downpour. Where twenty-four hours before we would have killed for this water, we now lay soaked, miserable and cursing in the cold. By sunrise we had slept little and now faced another formidable obstacle.

Snake Canyon was quite magnificent. More than two thousand feet deep and winding away to join the distant Colorado river, it formed a huge chasm at our feet. Five separate ropes, more than six hundred feet long, disappeared into the chasm from the canyon lip, dropping into a huge void down a wall that could not be seen even if you lay down and peered over the dizzying drop. We were above a huge, overhanging half-dome, and after donning climbing harnesses and helmets, we attached ourselves to the ropes and gingerly stepped back, down and out into space.

None of us had ever done anything on this scale before. The ropes had been specially made to rig this, the highest free abseil the world had ever seen. Helen, in particular, had had only limited training – abseiling inside a California gym – but in yet another demonstration that age need not limit your possibilities if your attitude does not, she stepped over the brink with a smile and a wave for the cameras. Then she raced down her rope faster than the rest of us.

The rappel delivered us on to a huge ledge part of the way down the canyon wall, and from there we traversed left to turn a steep corner. I went first and emerged on a small rock platform from which a rope spanned an enormous side canyon. It was about seventy yards long and pulled tight to an anchor point on a ridge beyond the deep cleft in front of us. That ridge was just a narrow blade of rock, and I could see a team ahead making their way along it, traversing the broken sandstone teeth. From their nervous and slow progress, it was obvious that another canyon lay on the far side, perhaps matching the thousand-foot drop at our feet. One thing was certain: this was not a course for the faint-hearted.

Helen came up behind me, exhaustion in every movement. The last few days had taken their toll, but as she reached the edge and stopped, she smiled as usual. There was no way that this was going to phase our granny. Casually she began to ready herself for the crossing, clipping into the belay point and examining the ropes that spanned the abyss. Then she sat down for a moment, obviously thinking about her situation. Up until now, Helen had only read about Tyrolean traverses, but here she was about to cross one of the most exposed spans I had ever seen. On the ridge ahead there were more of these rope bridges spanning the gaps between the huge rock teeth. There was no doubt that we were in the middle of some of the most dramatic rope lengths the climbing world had ever seen. While I checked her harness and safety kit, she quietly ate a powerbar and summoned more of that mental strength.

When I signalled that all was ready, she got to her feet, drew a deep breath and, on the count of three, jumped, and I pushed. She flew out over the void with a whoop of combined fear and glee, speeding down the rope with her pulley running free. Then, after perhaps a hundred feet, she began to slow as she reached and passed the mid-point of the span, eventually coming to a halt dangling right over the centre of the canyon. Now she would have to work to get herself up the other side – pulling hand over hand up-hill.

However hard you tension these ropes, your weight stretches and bows them, and moving beyond the middle is demanding. Helen began the job but the further she went, the steeper the rope became. I could see her fighting her fatigue. Although her legs were strong from years of running, she had never taken on challenges needing this sort of upper body fitness. It was telling, and her face showed the pain. If she did let go she would be back in the middle of the canyon and weaker than before. We all shouted encouragement.

'Go, Helen, go!'

At last her hand touched the far side and painfully, slowly, she managed to clip a karabiner to the anchor point. There she rested briefly before pulling herself in. She struggled to her feet and sat down with chest heaving. Fighting to regain both breath and emotions, she raised her hand and waved me on.

One by one, we all followed, and once established on the rock blade, we began to move along it – walking, crawling, and climbing. We remained attached to ropes at all times, but although this granted safety, they did not remove the overwhelming sense of height. I had climbed for much of my adult life but had never been in such an exposed position, and for people like Helen and Mary, who had never known climbing, it must have been totally overwhelming. The broken blade was also not the end of the rock course. From the last gigantic tooth that thrust out into a bend of Snake Canyon, we descended by a series of abseils surrounded by nothing but air. Altogether there were eleven more descents to reach the river floor below and so it was late in the day when we finally came to the canyon bottom and once again were swimming, wading and scrambling towards the next goal. We would rendezvous that night with our support team and then take to the water for much of the rest of the race. There were four more days in which to complete the course.

*

For our hunter–gatherer ancestors, inactivity at any age must have run counter to evolutionary development. Older members of groups may have become somewhat slower as they aged, but they could not have foregone an active lifestyle. If they had done so, they would simply have been left behind. Perhaps this is what did happen. When older individuals, past useful breeding age, needed a share of food and held things up, their families and fellows simply abandoned them. Yet such a cynical view runs counter to logic. The rejection of the elderly would not have assisted group effectiveness in a world where nothing was written down. All wisdom resided within the elders, for it was they who knew where water might be found in the once-every-thirty-year drought, or where the herds had gone when they failed to follow their usual migration routes. They would also have made invaluable baby-sitters, not for trips to the Stone-age entertainment complex but on a daily basis while the younger adults sought for food. Group survival must have depended to some extent on grandparents and on selection of genes conveying longevity.

There is a mistaken belief that it is impossible for nature to favour genes conveying an increased lifespan. It follows a simplistic interpretation of Darwinism, that natural selection works only on individuals. If that were the case, any mutation that gave benefit beyond breeding age could not become established through selective advantage, since only genes granting an individual greater personal breeding success could become more common. A gene improving lifespan beyond active breeding years would therefore fail. In reality, however, natural selection acts on groups as well as individuals. Throughout early history our ancestors would have lived in small bands based around extended families who shared many genes. A gene that made someone live longer might not lead to that man or woman having more progeny, but might still make his or her group more likely to survive. Since other group members were close relatives, they would be likely to be carrying the same gene and hence, in the end, natural selection could favour the longevity gene. Nevertheless, it could only do so if there were no costs. The survival of the elderly had to come without automatic physical weakness.

The selection of physical fitness must have occurred at all ages. Those who were more capable of going the distance to find the foods, and of moving fast when hunting or escaping the animals that turned on them, would have had a better chance of surviving at all times of life. Similarly, those who could remain strong when too hot, cope when too cold, or keep moving when short of food and water, would have been the ones to live. More important, they would have bred more successfully. Genes conveying resilience of all types must have been passed from generation to generation over hundreds of thousands, even millions of years. The result was inevitable. Through the passage of eons, most of the peoples of our ancestry were cut down by the blind sword of selection to leave a species that was supremely fit and adapted for survival in the wild environments of the Earth.

*

The white-water rafting was to take place on a section of the Colorado river known as Cataract Canyon, where serious rapids with Grade 5 wave trains, drops, stops, and rocks abound. Our inflatable rafts were very small compared to those used by commercial companies taking tourists down the canyon with guides. We too were going to have professional help for this risky part of our journey. The clock would be stopped for the time we spent on the white-water section of the river and then started again once we were below the major rapids, where the guides would leave us to paddle on down river to our exit point.

Even our team's expert guide, a young woman called Molly, expressed some concern about taking us down the rapids in a smaller boat than she had ever used before, especially with an amateur crew which included a 72-year-old great-grandmother. Still, gamely she pushed her reservations aside and set off with us. During the long journey down to the first of the rapids, she maintained her delightful sense of humour and taught us some paddling drills. We needed to recognise the commands she would shout when the real test came the following day. Instructions to paddle right . . . left . . . forwards . . . backwards . . . had to be obeyed without question, immediately and with every effort.

The night was spent on an island in the river, with plenty of time for sleep, although again we were disturbed by rain. In the middle of the night, we had to get up to create a shelter from our upturned raft supported by the paddles, but this worked well and, despite a downpour, we slept until 5 a.m. As we arose the rain stopped, and the morning became clear. The appearance of the sun lightened the tensions we felt, and spirits were high as we cast off.

Around us were several other teams who had spent the night nearby. A few were now so far ahead that they had run the rapids before sundown the previous evening. With little paddling, we were all swept downstream by the increasingly powerful flow of the huge river. Emerald green banks swept by, contrasting harshly with the vertiginous, red-rock walls behind, but slowly they became smaller and then disappeared altogether. The canyon had narrowed as the walls closed in, plunging from the sky above directly into the speeding waters. The boats moved faster and the flow became a flood, with white foam flying. A distant roar echoed from the walls ahead, becoming steadily more insistent. Suspense rather than the volume of sound stifled our chatter.

Soon we were moving through the fifteen-mile rapids at great speed. Up in the bow, Helen was hanging on with both hands, jamming her shoes where the rubber walls met the floor to keep herself stable. Behind her, we sat two on each side with our feet hooked under rope lines that were rigged across the floor to keep us in-board as repeatedly our boat was engulfed in water. Molly sat as high as possible on the back of the raft, to enable her to take a safe rôute ahead. Every second or two she issued a taut command, steering deftly between the rocks and drops with a combination of our propulsive efforts and short thrusts from her own larger broad-bladed paddle.

Steering is a loose term for what can be achieved in a fourteen-foot raft in a maelstrom. With the river pouring hundreds of thousands of gallons a second through the narrow gorge – seemingly much of that into our boat – it was essential for us to avoid the worst of the waterfalls, rocks, or tree trunks jammed across the river. Manoeuvres were limited to turning the boat and paddling with maximum effort towards one bank or the other, as we hurtled sideways towards the next danger. All the time, we were trying to gain a part of the stream that would not lead to destruction. Just before a drop or a wave train was entered, the boat had to be turned bow forwards to ride the violence without over-turning. As we did so, we had to continue our frantic paddling to ensure that we were not caught in vicious undertows or giant whirlpools. Throughout, our tiny craft was more under than on the water. Every torrent-filled drop made me wonder if we could make it, but each time we popped up successfully. With every fall. Helen gave a whoop of approval – a delightful sound soaring above the roar of water but foreshortened as she disappeared beneath another icy green wave. It was a wet roller-coaster gone mad – a ride that made the best of Disneyland pale in comparison.

Around us, many of the other boats were less successful. More than half of them capsized or lost crew members overboard, and soon the river was full of people travelling rapidly with no transport while out of reach swept their empty boats turned-turtle. Between the rapids were occasional eddies in which to stop and rest, and here overturned boats could soon be sorted out, although not necessarily with every crew member back on board. Some of the boatless travellers failed to make the eddies, despite their best efforts at swimming, and on they swept down long sections of huge rapids in freestyle. It must have been an extraordinarily exciting but dangerous experience. Fortunately, no one was hurt.

At last we came out at the end of the cataracts. It was early afternoon and our arms were wiped out from their efforts. With two days to complete, those arms had much more yet to do. The remaining segments of the race would all depend on upper body power. Just to exit the canyon, we would need to climb thousands of feet of sheer cliffs, and beyond that there would be a long canoe trip. Worse, before we even reached the roped climb out, there were 40 miles of slow and sluggish river to navigate. Though a little peaceful tranquillity was appealing, a fourteen-foot inflatable would not be my choice of craft for paddling long distances.

Molly left us to the task, and with the race clock re-started we paddled through the remainder of the afternoon and on into the night without stopping once until we finally reached the exit point at 3 a.m. Here the darkness forced a stop – welcome for shoulders, arms and blistered hands, but all too short. At 6 a.m. a rosy glow caught the upper walls of the canyon and signalled that it was safe to start the climb.

Standing at the foot of the cliff was both frightening and awe-inspiring. It rose in a series of buttresses to disappear from sight far above, the thin climbing ropes connected to the rocks by tiny steel pegs. Even from below, it was obvious that while some of the ropes were long enough to span whole sections of rock, with ledges making a connection to the next rope, others crossed unbroken vertical or overhanging sections that were far too long for single rope lengths. There the rope-to-rope transfers would have to be made while hanging free. It would be utterly exhausting for the climbers among us, let alone for Helen and Mary. There were no doubts in my mind. We had come far, but this challenge would be beyond us, Whatever her determination, Helen could not scale this wall.

Through every moment of that long climb, I thought Helen would fail. It took me nearly two and a half hours to make the ascent, and by the time I reached the top my strength was decimated. Far below, Helen's distinctive helmet was just a dot, even through binoculars. Whit was climbing with her and carrying most of her load while patiently guiding and giving advice. He could not help her physically, and every foot of her ascent was an enormous strain on her desperately thin arms. Yet her will would not be broken, and inch by painful inch, she gained height as the sun crossed the sky and blazed down upon her slowly moving figure – a human spider crawling up a gigantic web.

Bill and Mary were with me, looking down as Helen drew closer. It was agonising to watch her changing from one rope to the next, with all her weight hanging from her thin arms. Clutching two jumars – the friction devices that slide up a rope but not back down – she climbed slowly. Arm up . . . heave . . . leg up . . . push, and repeat. When she came to a peg that connected the ropes to the rock, she had to clip in to a tape loop attached to it, by using a karabiner joined to her harness. It entailed pulling herself completely upright with just one arm while reaching for the loop with the other arm in which she held the karabiner. It was vital to clip into the peg loop with a short line so that the connection to your climbing harness would then take all your weight. Only then could she release the jumars which held her so that they could be re-attached to the rope above the peg. Time and again, Helen would fail at this exhausting manoeuvre. Either the strength would go from her pulling arm or her fingers would prove too weak to open the safety gate on the karabiner. She would reach up, try to clip in, and then fall back several feet with a violent jolt. Then, after a short rest, she would climb up and try once more, only for the process to repeat itself. It was pitiful to witness, and the strain showed in every line of her face. At the same time, to see this small old woman dangling from ropes on the gigantic rock wall was inspiring. Everybody added their will to her own as she pushed onwards, but with the sun's heat adding dehydration to her weakness, she became ever slower as three and four hours went by.

In the end, I was to be proved incorrect. My conviction that this was a task beyond even Helen's capability did not take account of the power of will. Five hours passed and Helen was still going. The sun reached its zenith – six hours on a baking wall with no water before she came within a few hundred feet of the top. We could actually hear her cursing at herself in most ungrandmotherly language each time she failed and fell back, but she would not give up. Then, at shortly before three o'clock, she finally pulled herself over the lip and lay quietly face down at the edge of the canyon. For a few minutes we let her be to release her tears. Then we picked her up and hugged her before continuing on our way.

*

What physical differences separate us, the modern humans, from our early roaming ancestors? Many would imagine there is a large gulf between us, but I contend that it is closer to a crack. Hunter-gathering lasted until about ten thousand years ago and in many parts of the world went on until much more recently. With no fundamental change in lifestyles, selection would have continued to generate men and women who were perfectly suited for physical activity and hardship. Furthermore, underlying these capabilities, people would also have been perfectly adjusted at the biochemical level, with harmony between their lifestyle, diet and health.

Can we say the same for modern societies? Ten thousand years is not long in terms of the evolutionary clock and so there has been virtually no time for significant genetic change. Yet civilisation has led to huge alterations in our environment which have put us out of step with our evolutionary background. There is now a discrepancy between our biological design and what we do and eat. It is a discrepancy with considerable implications.

On the positive side, it can explain phenomena like Helen. Admittedly she may be at the extreme end of a biological distribution of fitness but nevertheless she illustrates our potential. I believe that most of us have an astonishing natural strength which includes the ability to walk and run considerable distances every day, and we also have the capacity to survive in conditions that are scarcely imaginable.

On the negative side, the discrepancy between our lifestyle and our evolutionary design is at the root of many of our problems. The advent of civilisation brought with it a radical change from our ancestors' diet of fruits, roots, vegetables and grubs. With improved tools and more success with hunting, the proportion of meat in our diet must have expanded a little over the last 100,000 years, but it would not have been until the successful domestication of animals around 10,000 years ago that the consumption of meat and dairy products increased rapidly. The speed of change would have been too fast for our biological systems to adapt, and so it is fair to assume that the changes must have upset a number of our metabolic controls. Furthermore, with the farming of animals and planting of crops, we lost many of the uncertainties in food procurement.

A reliable food supply has many implications for health. Obviously it is beneficial in avoiding death or illness due to starvation, but there are some disadvantages when the supply becomes too good. Our bodies are genetically tuned for constant activity and set up to face a predominantly vegetarian and varied intake. They expect to meet seasonal shortages and frequent famines which no longer appear. As a consequence, our health and our hunger control mechanisms are deeply confused. Ten thousand years ago there would have been no fat people around, and problems which now dominate our hospitals – heart disease, strokes, diabetes and even some types of cancer – would have been rare.

*

We left the top of the canyon in orange evening light, heading for our last support team rendezvous. It was a simple ten-mile run, downhill on tracks and narrow roads. This was a delight. Although our arms and backs were tired after the paddling and the climb, our legs were fresh, for it had been nearly three days since we had walked. We could happily trot and relax, admiring the glorious scenery that surrounded us. For the first time I noticed the enormous numbers of blooms in the desert terrain – multi-coloured bells, as delicate as any you might find in a flowered English meadow, mixed with spectacular funnels of orange, red and pink bursting from harsh-spined cacti. The paths weaved their way between them, giving the appearance that they had all been laid out deliberately – the desert equivalent of a Japanese garden.

It was sundown when we reached the last stage-point which lay on the banks of Lake Powell, an enormous artificial expanse of water, formed from the damming of this part of the upper Colorado canyon system. As darkness fell we sat overlooking the placid waters and ate a good meal. Beside us, loaded and ready to go, were our two Canadian-style canoes. Mary and Whit would paddle one of them (Helen installed between to get some rest) while Bill and I would take the other with all our kit, food and drink. By the time we were ready to go, it was pitch dark, and with the moonrise still some hours away, navigation would not be easy.

The map showed that we were in a broad flooded side canyon which we had to follow for seven miles before joining the main lake. Once there, it would be fairly easy to maintain the right line, for the course was dead straight for another twenty miles or so and all we had to do was to follow a compass bearing through the night. Beyond that the canyon wound again but, once there, we would have the moon and then dawn to light the way. The difficulty would be finding the main lake in the first place. The side canyon had its own branches which ran in different directions. It would be easy to go up any of these by mistake, thinking it was a bend in our course. There was no current to guide us and the branches were much the same width as our winding route.

No sooner had we started than another problem became evident. The darkness was so intense that hills were a mere extra blackness against starlight. It was impossible to judge their size or distance and we had little idea how fast we were moving. Sometimes we would fix on a distant hill and make it our target for the next hour's paddling, only to find it was a big rock a hundred yards away. At other times we took evasive action to round a rock that turned out to be a distant hill. We tried to follow the map, noting every change in bearing so that we could mark it as the turning of another corner, but soon we became deeply confused and only learned of an error after we had taken a heading that was impossible for our course. Then we would turn round to try again. Once or twice I am sure we covered the same section of water several times in different directions, and in the end it took more than seven hours' paddling to arrive on the main lake at about 3 a.m. We had slept for less than two hours in the previous forty-five, and for the rest of the time we had worked with arms and backs. We still had more than thirty miles of calm water to paddle and a stiff wind began to blow against us.

The moon rose as we reached the main lake. It helped to confirm our position, and the rest of the night was simply long and fatiguing. Our bodies now protested bitterly, despite the rhythmic paddling putting one into a kind of trance. Every so often the trance overcame the pain and eyes would close as minds drifted away. The dream would then be broken by waking with a jerk and realising where we were. It all reached a climax at around 5.30 a.m. when my boat suddenly lurched to one side and almost overturned. At the same time, there was a splash behind me. Bill had gone overboard. He had fallen asleep and dropped off the canoe.

We pulled him from the water and stopped on the bank for some food. The sky was beginning to lighten and one by one the stars went out. Slowly the tops of the high rock walls changed from black through red to gold. It was the beginning of our last day of effort, and we knew now that we would complete the course. There were fewer than twenty miles of paddling before it would all be over. Refreshed in spirit if not in strength, we returned to our boats and went on.

Four hours later, with no further stops, we came to the line. We were not the first by any means. Indeed some teams had arrived more than twenty-four hours previously and another sixteen teams had made it in the interval. Still, at eighteenth out of fifty teams we, with our great-grandmother, were elated. The welcome we received said it all. As we disembarked from our canoes loud cheers greeted Helen. For her, it was an extraordinary moment of triumph.

TWO

★

Out of Africa

IN ORDER to understand fully the origins of our resilience and the problems that arise from conflicts between biology and lifestyle, we must examine our evolution more closely. For hundreds of years it has been clear to many thinkers that man is one of the animals. Even by the second century A.D., the Greek physician Galen had dissected various creatures and recorded that human beings are most closely related to the monkeys. It was obvious to him that our bones are similar in shape to theirs, as are our organs and muscles. Since then, the reason for this relationship has become clear. The emergence of Darwin's theory of evolution explained how different families of animals came into being and now, a century and a half later, the evidence is conclusive. We humans are full members of the sub-group of primates known as the apes.

The apes include gorillas, chimpanzees, orang-utans and gibbons, and at different times ideas have varied as to which of these is our nearest relative. Today a majority of the public believes that this dubious honour must go to the gorilla, although not long ago some experts would have advocated the orang-utan. New techniques, however, have shown that neither candidate fits the bill. Studies comparing our DNA with those of our primate relatives have provided an unequivocal answer – the chimpanzees are our closest cousins.

Some six or seven million years ago, in Africa, there was a population of small apes that were the common ancestors of both man and the chimp. The population must have then divided into two groups which started to live apart and tended not to interbreed. The cause of this division is not entirely clear but the idea most popular among anthropologists is that the two groups started to occupy different environmental niches. At around that time the African climate became both hotter and drier, and it seems that this led to a thinning of the forests and an increase in the areas of savannah. With a reduction in the number of trees, some of the apes started to live on the ground and then moved out on to newly open grasslands. Others remained more arboreal, living within the forests. The result was that the two groups were exposed to different environmental demands and this led to their slow evolutionary divergence. The tree dwelling group slowly evolved towards the two species of modern chimps. The ground dwellers embarked on the path that would eventually lead to man.

One of the changes that life on the savannah appears to have prompted was a move to becoming upright, and various suggestions have been made to explain this. An early interpretation was that the change occurred in order to free our hands for tool use but the fossil evidence does not support this. The change was already well established before the first evidence of rudimentary tool-making and expansion of brain size four million years ago. Another suggestion is that the move to the savannah meant that the groups had less access to shade. With the sun high in the African sky, heat stress would have been a considerable problem and the adoption of an upright stance would have reduced exposure by limiting the surface area under direct rays from the sun. The same radiant heat argument has also been used to explain why we became bare-skinned except for the tops of our heads, but if the lack of shade was the driver towards naked uprightness, it seems odd that other species, living in equally hot and sunny circumstances, have not found the same necessity for change. Another theory suggests the erect posture was adopted in order to reach fruits in the smaller trees and bushes that were growing in the new hot, dry weather and were too frail to be climbed. These ideas have led to the arguments being reversed. Instead of the use of tools bringing us to our feet, it seems more likely that it was the coming to our feet that promoted the use of tools, which in turn drove the rapid development of brain size and intellect – features that were to become the greatest assets of humanity.

It might be thought that an increase in intelligence would confer considerable survival benefits for any animal but, without simple tools, intelligence may be less of a survival advantage than one imagines. Even if you are a lion with the capacity to understand astrophysics, your advantage over a more stupid lion is fairly limited. Your knowledge would be of little help in mating or fighting, and although your generally increased intellect and canniness would give you occasional benefits on the hunt, it would not necessarily make you more successful than rivals who were faster or fitter. It therefore seems likely that selection for intelligence exerted only a modest effect on the development of most species and that many modern animals are little more intelligent than their ancestors. At the same time, increased intelligence does promote considerable survival advantages once it can utilise technology. Once our ancestors were upright, the freeing of their hands and the consequent use of tools made intellect a much more powerful asset, and the chances of survival for brighter members of a group increased considerably. The evolution of intelligence and technology must have accelerated together.

Although the divergence in intelligence and posture between the ancestral chimps and humans is striking, it would not have happened quickly. Mutations are infrequent and the majority of them fail to become established in populations. They arise from mistakes in the copying of the DNA code as it is passed from one animal to its offspring but many occur in parts of the DNA that are redundant. These result in no change of either structure or function in the animal that inherits them. Transmission to future generations is entirely a matter of chance, and with all the other animals of the group passing on the 'normal' form of the DNA, most mutated sequences in non-critical regions of DNA fail to become established in the population.

Of course some mutations do occur in a region of DNA that *is* active, and there is a change in a gene that actually encodes one of the animal's proteins. All proteins are made up of long chains of linked amino acids and the mutation causes the substitution of the wrong amino acid somewhere in the chain. This tends to alter the protein's shape, normally critical for its function. Most proteins either contribute to an animal's structure or operate as an enzyme to control one of its metabolic systems. In either case, their exact shape is usually crucial for their role. In fact the amino acid sequence is close to being inviolable, and because of this, most mutations in critical DNA regions are immediately fatal. They also go nowhere.

Only in rare instances does a mutation lead to an amino acid substitution that actually improves a protein's function. If this occurs, natural selection can come into play to increase the chances of that mutation becoming widespread in the population. Nevertheless, it must be remembered that the new mutation will become universal only if the descendants of the animal that first had it are the sole survivors. This happens either through exceptional random chance or when the advantage conferred by the mutation is really very considerable. Many advantageous mutations fall by the wayside and the overall pace of evolutionary change is slow.

At first sight, it might be imagined that there were large genetic differences between ourselves and the chimpanzees, perhaps suggesting that the acquisition of new mutations was somewhat easier than I have intimated. This is not the case. In reality, the differences between our genetic codes are tiny and direct comparisons have demonstrated that 98.4 per cent of chimpanzee and human DNA are identical. This is less of a difference than that seen between many animals designated as being of the same genus, and some anthropologists believe that it is mere pride that stops us redefining our position in phylogeny. In *The Rise and Fall of the Third Chimpanzee*, Jared Diamond points out that we are actually another species of chimp rather than a member of a separate grouping. Certainly we are more related to the chimps than they are to gorillas.

Despite many areas of debate, the steps that occurred during the period that separates us from the chimps are becoming clearer. By three million years ago, our upright ancestors had divided into at least two different lines, with several species in each line. There were heavily built versions of early humanoids of which the best known is *Australopithecus Robustus*. They survived for a very long period with little obvious development of either anatomy or technology but eventually their luck ran out, and they were eliminated by rivals more than a million years ago. The lighter built versions fared better. The best known of these is *Australopithecus Africanus*, but there were many others, differing only slightly. All were bi-pedal. They lived in various locations in Africa, and by two and a half million years ago one group had evolved large enough brains to allow the scientists of today to place them in the human genus. They are known as *Homo Habilis*.

When *Homo Habilis* appeared in the fossil record all early humans were confined to the African continent. They made tools that were probably good enough to give them a real advantage, and with that their development accelerated. By around 1.7 million years ago, they had evolved further into *Homo Erectus*, which had an even larger brain and presumably greater intelligence. Perhaps it was this that granted them the flexibility and imagination to move, and around one million years ago *Homo Erectus* began to spread and became established in other regions of the globe. It would be easy to imagine that these populations were the forerunners of the diverse races of today but, although there is considerable debate about the point, with claims that Australian Aboriginals are descended directly from *Homo Erectus,* it is probable that the ancestors of all current humans did not leave the African cradle until very much later. The confusion arises because, oddly, the precise history of our ancestors in the last million years is more difficult to interpret than that of earlier times. The later groups wandered much more widely than their predecessors, creating considerable confusion in the fossil record. However, a group that left Africa in a migration around 100,000 years ago was very much closer to all modern races.

Following this second African exodus, three lineages of *Homo* seem to have inhabited different parts of the world. One, in particular, fits the popular conception of early cave men. These were the Neanderthals – men and women who must have looked quite different to ourselves, with muscular bodies, flattened foreheads, heavy bony eye-brow ridges and faces that were prominent. Around 100,000 years ago, they were widespread and their remains have been found in caves over much of Eurasia. It is likely that most Neanderthalers actually lived in temporary shelters and that their discovery in caves is merely a quirk of preservation. Nevertheless, it is doubtful that they did much in the way of complicated construction. Although their brains were large, indeed larger than our own, they appear to have been very inferior intellectually. They made only primitive tools and had little or no art or culture. A long way from modern man, they finally disappeared around 40,000 years ago. Quite why remains a matter for speculation, but it seems most likely that they were in direct competition with other species. It is certainly possible that their extinction was due to our own ancestors embarking on the road to genocide.

Another of the three human species that lived 100,000 years ago was limited to parts of Asia. Examples have been found in the caves of Java and elsewhere in the Far East, but relatively few remains are known and so the development of their anatomy, tools and culture is not clear. They too were descendants of the first *Homo Erectus* migration but, like the Neanderthalers, they died out and left no descendants.

It is the third type of early human that holds the key to our modern existence. These are the Cro-Magnons, first recorded in African sites of around 130,000 years ago. They were very widespread on that continent, with the best fossil records found in the caves of South Africa. Around 100,000 years ago, it was some of the Cro-Magnons that undertook the second dispersal from Africa with the result that their remains began to appear over much of Eurasia, and later groups moved to the Americas and Australasia. The Cro-Magnon remains present a picture of advanced tool-makers who must have looked very like us. They hunted successfully, albeit for quite small prey, and must have gathered the remainder of their diet. Early in their history they did not display much in the way of advanced culture, but around 40,000 years ago this changed quite suddenly.

There is considerable speculation as to the trigger for the acceleration in Cro-Magnon development. The explosion seems to have been cultural rather than physical, for there are no simultaneous changes in their fossil records. One possibility is the discovery of language. Although it has been shown that groups of wild chimps use a limited vocabulary of calls to co-ordinate actions such as hunting, their capacity for communication is very restricted. In part, this can be ascribed to their inferior intelligence, although studies in the last few decades have shown that domesticated chimpanzees can use signing or other techniques to represent items or actions separate in time or space. They therefore possess an intellect adequate for some language, but if sounds are to be used to represent concepts, a wide range of vocalisation is required. When compared with humans, chimpanzees are restricted in this respect, due to the different anatomy of the mouth, pharynx and larynx. It is probably these differences that have limited the development of the chimps' communication.

Unfortunately, anatomical differences dictating vocal abilities leave no fossil imprint, and so we cannot define the time at which early humans developed speech. Nevertheless, we can surmise that once this did occur it would have precipitated a marked step forward – stimulating the further selection of intellect in the same sort of way that freeing of the hands and the use of tools had done. With no change in underlying brain complexity or innate intelligence, language would have launched a new era of social interaction and group success. Was it therefore speech that launched Cro-Magnon on the path to modern man?

How different are we from the Cro-Magnons or, more important, from their hunter-gathering descendants of immediate pre-civilisation? The answer is important, for if we are genetically very similar, it would support the supposition that we are designed for a lifestyle at odds with our own. Evidence can be gained from the DNA. The fact that we share 98.4 per cent of our genes with the chimpanzees means that only 1.6 per cent of our genes must account for all the obvious differences between us – intelligence, erect posture and speech. The remainder of our physiology and metabolism must therefore be very alike, and indeed this is the case. Biochemically, we could not be much closer cousins. Since the fossil record shows that the joint ancestors of chimps and man lived between five and seven million years ago, it must have taken around three million years for the DNA of the two groups to diverge by just one per cent through the acquisition of new mutations. There is no reason to assume that this rate of DNA change is not typical for all primate or hominid groups through the ages, and so the 40,000 years that separate modern humans from the mid-Cro-Magnon years represents a genetic difference of little more than 0.1 per cent – a difference within the range of variation that exists between modern human races.

This type of speculation, indeed my earlier description of slow change through the blind selection of mutations, can be viewed as overly simplistic. In his recent book, *Lifelines,* Steven Rose argues that the reality is far more complex and that evolution acts on complete individuals within their environmental framework rather than by a mechanistic, constant 'selfish gene' process as advocated by Richard Dawkins. Rose may be right and, if he is, then changes in our genes could have occurred somewhat faster in the last 40,000 years, when rapid development of intelligence may have pressured the selection process. Still, I doubt if it could have made much difference to the overall rate of biological adaptation and, as I have mentioned previously, it was only ten to fifteen thousand years ago that modern man moved away from Cro-Magnon life and began the domestication of cattle and the setting up of more permanent homesteads. The genetic differences between modern humans and those early neolithic farmers must therefore be tiny, probably less than 0.05 per cent, and so if it were possible to extract late hunter-gatherers from the past, I believe that they too could be trained as fighter pilots, doctors and lawyers. Of course, the converse is also true. We, the pilots, doctors and lawyers of today are almost identical to those men and women who dwelt in caves and shelters. We are born superbly designed and equipped for action and survival in a harsh world – but unfortunately not for the one in which we live.

THREE

★

A Gene for the Fastest

W HEN Linford Christie or Donovan Bailey go down on the blocks for the sprint to glory or dismay, they command bodies that are designed for moving rapidly. Since birth their genes have dictated that their leg muscles would be packed with a specialised, fast-contracting type of fibre, but that endowment has not been enough. Like all world-class athletes, they have put in endless training to go so far, building upon their birthright to improve their muscles' size, strength and short-term fuel supplies. Their training incidentally also honed their hearts and lungs to the point where they can now pump round twice as much oxygen as can an average man's. Yet, even boosted by a pre-race priming of adrenaline, heightened by the expectancy of a hushed stadium, it is not this exceptional oxygen delivery that will make the difference. The athletes will work too hard and the encounter will be too brief to use their aerobic fuel systems. They are the ultimate speed machines, with legs set up to produce a short burst of power that is dependant solely upon the chemical energy stored and ready within them. As far as those legs are concerned, the whole race could be run were their hearts to stop with the shot from the starting pistol. Their brains, of course, would be less flexible.

The gun is fired and a single pressure wave of sound is received by the competitors and passed, almost by reflex, to their motor cortex. There it triggers the conscious drivers of movement that hit the neural switches of the 'go' command which then stamp on the accelerators. Muscle engines explode into life as more than one hundred thousand fibres in each limb hurl the competitors forward. The sheer power of each of their huge muscle groups is then blended into a perfection of movement that can only be appreciated in the later slow-motion video. They burst from the blocks with the balletic grace of a Nureyev but the acceleration of a Ferrari.

Once away and into their punching run, most of the control is switched to the auto-pilots of the lower brain-centres programmed with patterns of muscle firings burned in by experience. The learning goes right back to the first tottering steps of their childhood, but unlike most of mankind, who simply added the everyday experiences of walking and running to such infant movements, their training has led to perfection. They have spent days, weeks, years to reach this level, and to the extraordinary power of their muscles has been added deliberate assessments of running style. With effort written in near frozen expressions of will, the athletes are now capable of the ultimate sprinting achievement – winning the Olympic hundred-metre Gold.

★

Although the high levels of activity undertaken by our ancestors led to the selection of men and women who were all physically fit, it did not confer the same type of athletic ability on every individual. Athletes of today cover the spectrum between those designed for short-term speed or power events, such as sprinting or weight lifting, and those designed for endurance events, such as the mile or the marathon. Underlying these specialist capabilities are variations in physiology which have only recently been understood.

All animals are designed for movement. As evolution led to larger and more complex organisms, they could no longer remain reliant on the limited food sources that lay beside them or washed slowly by in the primitive seas. They needed to travel from place to place to find more to eat, and for that they needed a means of locomotion. Over millions of years, all the higher genera developed a system based on muscles – specialised contractile organs that could propel them around the waters, the land or the air. In the primates, those muscles and the skeletons through which they operated, became the greater part of the body. Today nearly three quarters of our weight is dedicated to movement.

A useful analogy is to compare our locomotion systems with those that drive modern vehicles. In cars or trucks the energy for movement is created by burning fuel in the cylinders and then utilising the expanding explosive power to move a system of levers, then cogs, to turn the wheels. At the same time new fuel and air are delivered by fuel pump and carburettors and the waste products of burning – carbon dioxide, water and a few more noxious chemicals – are cleared away through an exhaust system. Excess heat is lost through the cooling system.

In man, as in all other large animals, the single combustion engine is replaced by numerous separate muscles which act like servo-motors seen in more complex man-made machinery. As with internal combustion engines, each muscle runs by combining fuel with air, although they can also run for short periods off chemical storage of energy. Instead of the inefficient explosion, the burning is much better controlled and the muscles produce a contraction power stroke rather than an explosive outward one. The power released is then harnessed by a series of levers which grant us motion directly rather than it being transformed into rotation. Evolution has never devised animals on wheels, probably because all parts of an organism need to have a blood supply and, unlike electricity, blood cannot be passed through brushes to or from a rotating axle. It is indeed difficult to conceive of how a living, wheeled locomotion system might have evolved. Furthermore, wheels have no advantage on uneven surfaces, and only in very recent times has our intellect developed sufficiently to allow us to smooth the earth's surface at will. Now, joined to inorganic wheels, the man or woman on a bicycle is probably the most efficient motion machine the world has ever seen.

Altogether we have around two hundred and fifteen pairs of muscles arranged symmetrically down right and left hand sides of our bodies. Most of these move bones at joints around which there are usually several different muscles with varied actions. The movement of each thumb, for example, is controlled by nine different muscles whose functions are complementary, acting to assist one another. Complementary does not mean that they necessarily move the joints in the same direction; in fact, every joint must have muscles set up to work in opposition since, as a muscle can only pull and never push, one could never reverse a movement without such an arrangement. Opposing muscles also allow smooth movement by tensioning and controlling both sides of an action, and they grant joints increased stability.

When, in 1660, the Dutchman, Anton van Lewenhooek, looked down one of the world's earliest microscopes, he saw that muscles are made up of tens of thousands of elongated fibres. Each of these is a single cell, with some of them among the largest cells in nature. Even in man, only a medium-sized animal, one fibre can be more than 30 centimetres long, running the entire length of large muscles such as those in the thigh. Yet, despite their length, these fibre cells are only a few thousandths of a millimetre thick. They therefore have the shape of an extraordinarily thin cylinder. Although each muscle fibre is actually capable of independent contraction, in reality they are covered by fine fibrous tissue that joins them into bundles. These bundles are then further linked to form the muscles themselves, with the fibrous coverings blending to form their tendons. As a result, thousands of fibres can pull on the skeletal levers simultaneously and can generate extremely large forces. These have led to extraordinary attainments. Individuals have cleared high jump bars 60 centimetres above their heads and pole vaulted well over 6 metres; they have run 100 metres in 9.8 seconds and 100 miles in eleven and a half hours; they can score 501 on a dart board with just 9 darts and achieve maximum breaks of 147 on the snooker table; they have even generated a force equivalent to 442 kilogrammes with a single bite. We may not be the fastest or strongest animals in the world but we are surely the most adaptable.

*

Inside each muscle fibre there are the structures common to all living cells – the systems that fuel, repair and re-build them as necessary – along with the specialised contraction mechanisms. The contraction systems lie within fibrils, even tinier cylinders that run the length of each fibre in a manner similar to that in which the fibres run the length of muscles. Fibrils can also contract independently if artificially isolated, but it took the invention of the electron microscope to understand how this occurs.

When your brain wishes to move a muscle, it transmits electrical impulses down the nerves of the spine and out along controlling nerve branches where they pass along the surface membranes of the muscle fibres. These membranes are folded in such a way that the electrical signal is carried close to all of the fibrils within. Each fibril is made up of chains of thousands of tiny contractile units which consist of two end plates with two different types of protein filaments connecting them. Actin filaments are very thin and extend from attachment points on each of the end plates towards the central part of the unit where they lie adjacent to one another, and in parallel. In the central region the thin actin filaments are interspersed with thicker filaments of myosin in a perfectly regular pattern so that each end of any myosin filament is surrounded by six actin ones extending in from opposite end plates.

When the muscle is relaxed, there is no contact made between the actin and myosin although hundreds of tiny extensions come off the side of the thicker myosin and almost touch the actin ones. It is these molecular-sized arms that are the key to movement. On the arrival of the nerve impulse, calcium is released around them from specialised microscopic holding tanks. It has two dramatic effects. First, it makes the tips of the tiny myosin arms sticky, so that they attach to the nearest actin filament, and then it makes the arms deform and bend, as if hinged, at the point where they extend from the main myosin filament. Because the arms at each end of the myosin bend towards the middle, the stick and swing action pulls the surrounding actin filaments fractionally in towards the centre. With them come their attached end plates and so the electrical impulse has resulted in a shortening of each unit.

Of course the shortening of an individual contractile unit in response to one nerve signal is too small and weak to be of any use on its own, but it occurs in thousands of units linked together in each fibril, in thousands of fibrils that make up each fibre, and in thousands of fibres that make up each muscle. The single electrical trigger can therefore result in a powerful muscle twitch. Yet muscle twitches alone, however strong, are also useless when it comes to controlled movement. It is necessary for muscles to shorten steadily, and this is achieved by repeating the triggering process rapidly. No sooner have the myosin arms stuck, bent and pulled, than they release again as the calcium is quickly pumped back into the storage tanks. Then a second electrical signal arrives before there has been any time for things to relax, and a new stick and pull cycle is initiated. With the brain capable of triggering 100,000 shortenings per second, the result can be dramatic. The muscle exerts a steady shortening pull, and by regulating the frequency of firings and the number of fibres triggered, the brain can exercise precise command over considerable forces.

The mechanisms that allow muscles to contract need a supply of energy to do their work, and this is achieved in the same way as the body fuels all processes of life – from the breakdown of chemical bonds within compounds. All chemical bonds contain energy but we can only use the types found in foods such as carbohydrates, fats and proteins. The energy in these comes originally from photosynthesis in plants – the process powered by radiation from the sun in which leaf chlorophyll bonds carbon dioxide with water to make sugars.

Between the plants using sunlight to trap energy in sugars and our utilising that energy for action there are a series of pathways. The first set of pathways lie outside our bodies and can be either straightforward or complex. In some cases we simply eat the plants and so ingest the sugars directly into our system. In others, however, the plants themselves convert sugars into different substances, or we eat animals that have consumed the plants and made other conversions. Overall, we ingest sunshine energy converted into a whole variety of chemical bonds in proteins, fats or more complex carbohydrates – molecules which then enter a second set of chemical pathways inside our bodies which are also of varied complexity.

At their most simple, the sugars or fats are absorbed and transported directly to muscle cells where the chemical bonds are broken and the energy is released and used for movement. More often, however, the energy from our food goes through a host of other intermediate reactions inside us, frequently spending time in either fat stores or body structures. Nevertheless, it eventually meets the same outcome. The sunshine that was incorporated into sugar bonds within the leaves of plants eventually runs our muscles as it does our other metabolism. As this happens, carbon dioxide and water are released once more. Indirectly, we are all solar powered.

The chemical energy within our food needs to be changed into a form that can be used at the molecular level in cells. Evolution has chosen a molecule known as adenosine triphosphate (ATP) for this purpose and it exists very widely in the body, fuelling the majority of cellular processes. ATP can provide energy conveniently because it is easily split by an enzyme to release a small packet of energy, and the splitting enzymes can be located wherever that energy is required. In muscle they form the very tip of the myosin arms and one molecule of ATP is used by each arm for each stick and pull cycle.

Muscle fibres come in two different types depending partly upon the nature of the ATP splitting enzyme. One type, fast fibres, can split ATP extremely quickly and so allow contractions of great speed, whereas the other type have a slower working system. There are, however, a number of other differences between fast and slow fibres.

When the gun fires at the start of a 100-metre sprint, the athletes' pre-made muscle stores of ATP can provide only enough energy to last for one or two seconds. In order to counter this, the muscles have a number of ways of replenishing the ATP supply. Fast fibres contain relatively greater stores of another simple substance, creatine phosphate, that will also split easily to release energy which can be used to regenerate ATP instantly. Yet even fast fibre stores of creatine phosphate are limited, only providing for ten to fifteen seconds of maximum effort, and although this is enough to decide the outcome of any 100-metre race, the levels start to decline from the moment the race commences, and if they were the only means of producing more ATP quickly, the runners would slow down well before the line. This does not happen. The creatine phosphate system is actually just a short-term back-up to give time for more complex energy production systems to come on line. These can regenerate much larger amounts of ATP for much longer periods and are of two types – anaerobic and aerobic.

Anaerobic metabolism generates energy to maintain muscle ATP without using oxygen – a huge advantage since there is no need to rev up the heart and lungs, before the process can go into action. The system can take over energy provision very quickly and so predominates in fast fibres. At the same time, using anaerobic systems has a downside. The energy is generated by the conversion of glucose into lactic acid, and the chemical bonds in lactic acid contain nearly as much energy as the glucose bonds in the first place. The process therefore releases very little energy from each glucose molecule and so glucose in the muscles is used up quickly. Although the glucose can then be replaced from a store in the form of glycogen, this takes time.

Glycogen consists of lots of glucose units linked in chains. When anaerobic work first starts, an enzyme splits the units off efficiently, so that the sprinters' muscles can go on with full anaerobic energy generation. But if high levels of work are sustained for more than a few seconds, the lactic acid formed from the process begins to accumulate. This poisons several enzyme systems, including the one splitting glucose from the glycogen. Within a 100-metre race, such lactic acid effects hardly matter, but go on for more than 200 metres, and even top class sprinters begin to falter. It is lactic acid build-up that explains why world records for 100 metres are well under ten seconds, for 200 metres are less than twenty, but for 400 metres are around 43 seconds and for 800 metres close to 100 seconds.

Slow fibres use predominantly aerobic pathways to generate energy. In these, glucose, fat or sometimes protein are 'burned' with oxygen to give carbon dioxide and water which are very simple molecules containing little chemical energy. Aerobic systems can therefore release far more energy than anaerobic ones, and since lactic acid is not produced, they can go on fuelling muscles for extended periods. Effective as they are for prolonged work, the rate of energy production is limited by delivery of oxygen by the heart and lungs and subsequent removal of carbon dioxide. They sustain only moderate work levels compared to the anaerobic systems.

The key to being a good sprint or power athlete is to have muscles with plenty of fast fibres in them, whereas the key to being good at endurance events is to have lots of slow fibres. The preponderance of one type of fibre or the other in your muscles is dictated by inheritance. It is your genes that make you go fast or far.

*

Although most anthropologists believe that our ancestors became erect to improve their reach, the requirement for speed has also been proposed as the primary reason beside those other theories of heat stress and tool use. This seems unlikely. Most four-legged animals run much faster than we do, and even chimpanzees and gorillas can set a fair pace with their knuckle walking or running. Nevertheless, when our ape ancestors first moved on to the savannah, they would have been more specialised brachiators than the large apes of today, with fore-limbs that were very useful for climbing but probably poor for aiding movement on the ground. It is possible that getting up on two legs could have conferred some speed advantage, and so the hypothesis must remain in frame.

Whether or not it influenced our posture, a turn of speed would have always had survival advantage. The faster our ancestors could run, the more successful they would have been at catching quarry or escaping predators. Yet, because of the specialised division between fast and slow fibres, and the consequence that absolute speed can only be gained at the cost of reduced endurance, the balance of advantage must have varied with local circumstances.

When most of us run hard, even for a few seconds, we rapidly feel our legs turning to jelly as the lactic acid builds up. Indeed, many of us experience this when doing no more than climbing a flight of stairs. Obviously we are not material to win an Olympic sprint medal, and although this is partly because we are untrained, it is also likely that we were never born for sprinting. Good sprinters may have running muscles consisting of as much as 90 per cent fast fibres whereas the average person has closer to a 50 per cent fast and 50 per cent slow mix. This gives most people a balance between speed and distance capabilities which must have been optimal through much of our evolutionary past. After all, it is no good being able to run like crazy for the few seconds needed to kill a quarry if you do not have the endurance to track it down in the first place.

If natural selection favoured a fast/slow mix for most people, why have some individuals ended up with muscles containing so many fast fibres? While the speed of DNA change means that the 100,000-year-old African Cro-Magnon population must have been almost identical to all modern humans, this does not mean all evolutionary development ceased during this period. After all, there are obvious racial differences today. However, since 100,000 years is but a short time in evolutionary terms, the only features likely to have changed much are those for which simple mutations of just one or two genes can create a quite definite survival advantage. An obvious example is colour.

Skin pigmentation is fairly simply inherited and strongly selected for by climate. No doubt all man's early African ancestors would have been dark, with melanin in the skin providing vital protection from excessive sun-induced damage. After men moved to cooler, less sunny regions, a high skin melanin would have had distinct disadvantages. Vitamin D production, needed for bone metabolism, relies upon processes activated in the skin by sunlight and, if the sunshine is weak and the skin dark, rates of synthesis can be seriously compromised. More northerly dwelling individuals would have had a considerable advantage if they were paler skinned, and so selection would have favoured the pale skin genes. Even so, rickets caused by vitamin D deficiency was common in north European races right up and into this century, and would have been commoner still if it had not been offset to some extent by the invention of farming around 10,000 years ago. Not all vitamin D is manufactured by body chemistry and some is obtained from meat and dairy products. Domestication of goats, sheep and cattle probably spared the northern dwellers many problems, although today they have arisen once more. With recent technology allowing more rapid migrations, vitamin D deficiency is now seen frequently in dark-skinned vegans who have moved to low sunshine climates with no time at all to adapt – especially among migrants to Britain from the Indian sub-continent. An opposite problem has also arisen for peoples with low skin melanin levels migrating at speed to areas of sunshine: skin cancer is increasingly found in white Australians and South Africans, and even among over-enthusiastic northern holiday makers.

Like skin colour, the predominant fibre type in muscles is a simply defined genetic feature for which local environments might pose strong selection pressures. It is particularly notable that a high proportion of Afro-Caribbeans are among successful world-class sprinters, and it is known that the average Afro-Caribbean has a considerably higher fast fibre count than other races. So it seems likely that their ancestors lived in situations where sprinting ability was a particular advantage compared to the need to run great distances.

The environment of West Africa, where Afro-Caribbeans originated, would have been a situation with plenty of food for gathering and plenty of animals that could be better hunted, or perhaps avoided, with a burst of speed. In these circumstances the advantages would lie in being able to run fast rather than far. In contrast, mountain environments, such as those of Ethiopia, have produced many of the world's greatest long distance athletes. In such regions there would have been far fewer animals to hunt and much less food to gather. The early men and women would have done better to move more slowly but have the capacity to rove far and wide in search of food.

Although differences in ancestral environment could underlie the current dominance of Afro-Caribbean runners in world sprinting, it is also possible that the selection pressures responsible were of more recent origin. Many of the current top class sprinters would have had ancestors transported by slave traders to the Caribbean or the United States around two hundred years ago. Those men and women may have been unwittingly selected for speed as opposed to endurance by ships' captains who deliberately picked the strongest-looking individuals, for muscularity is associated with high fast fibre counts. Furthermore, once committed to those inhuman transport ships, additional selection for strength might have taken place when the weaker succumbed to the extreme hardships, and even after reaching the plantations the same sort of factors may have operated.

Support for the significance of this new process of 'unnatural' selection is derived from the fact that it is the American and Caribbean populations of Negroes that do so well at speed events rather than the present populations of West Africa. At the same time, their different social situations may also have influenced the situation, and indeed the whole phenomenon of Afro-Caribbean sprinting success is arguably a social issue. With the prejudice that still exists in western societies against black-skinned people, success in sport is one of the few ways of rising from the bottom of the heap, and many experts believe that such a spur could easily account for all their sprinting success. To my mind, however, this simply does not fit the facts. There is plenty of prejudice against many other races in modern society, yet other minorities are not so successful in the world of sprinting. Furthermore, if Afro-Caribbean success was entirely socially stimulated, why have they been less successful at races over longer distances?

Of course, by discussing racial differences from a genetic point of view I tread on sensitive ground and open myself to accusations of Malthusian thinking. My argument, however, is not prejudicial against any race and I certainly give no credence to suggestions that there are genetic differences in intelligence between races. Unlike muscle fibre type, the genetic specification of intellect is extremely complicated, and so could only change very slowly. Once we became upright tool-users, any advance in intelligence would have been a huge advantage to all early human groups wherever they lived, and from this there could be only one possible outcome. All modern races of man must be essentially identical in intellectual capacity and, for that matter, were we to find a late Cro-Magnon in our midst, there would be no reason why he or she could not collect a Nobel prize.

*

Although I have dwelt upon the genetic background that is needed in order to become a world-class sprinter, I do not want to leave the impression that the majority of men and women are poor at it. Because most of us have around 50 per cent fast fibres we can all sprint, and it is only due to our rarely indulging the capability that we think of ourselves as being none too good at it. Successful sprinting is only possible if your natural endowment is developed through training, and the right training will have a spectacular effect on anybody's speed performance.

The most striking change seen with either power or endurance training is an increase in the size of the muscles involved in the activity. It was once thought that this was due to an increase in the number of fibres inside the muscles, but it is now known that this remains unchanged from birth and it is the size of individual fibres that increases, with very specific enlargement of the type of fibres most appropriate for the exercise demands being made. Sprint or power type of work will therefore enlarge the fast fibres, with an increase in the number of contractile fibrils within each one, and simultaneous improvements in the enzyme systems that provide anaerobic energy and the development of greater tolerance to lactic acid build-up. However, these anatomical and biochemical changes are not the only benefits. There are also dramatic changes in neuromuscular co-ordination which have far swifter and greater effects than the changes in muscle size and content.

Every muscle is supplied by nerves running from the spinal cord, each of which controls a group of muscle fibres known as a motor unit. The number of muscle cells in a unit varies with the degree of fine control needed by the brain. In the large muscles used for sprinting, 5,000 muscle fibres may be controlled by a single nerve fibre, whereas in the muscles that control the eye, the number is closer to five. While the brain can exert extremely fine control of eye movement, it has only much coarser control of running action. Even so, it can still regulate the speed and strength of leg contractions pretty accurately.

When you are walking, relatively few motor units – a few tens of thousands of muscle fibres – are being switched on and off in your leg muscles, while when you run, hundreds of thousands may be employed. The faster you try to go, the more units come into action, but if you are untrained you will not activate many that could be of help, and so you will fail to generate the maximum strength and power of which your muscles are capable. With training, the situation changes rapidly. After only a week or two of regular sprinting, many more motor units can be fired simultaneously, as if the muscle has learned to perform the action more efficiently.

I have noticed this phenomenon myself on several occasions while attempting to train for power fitness. By nature, I lean towards endurance events rather than speed, and endurance has been a common feature of my long distance Polar expeditions which have entailed dragging heavy sledges across Arctic or Antarctic ice. However, such journeys do require some high intensity anaerobic type of work, for sledges often become stuck on ice slopes and take brute strength to move them. To cope with this, I have needed to mix some power work with endurance training when preparing for these trips.

Unlike most professional athletes, I have never been good at maintaining levels of fitness in the periods between expeditions. Occasionally I go for a run, but I do no power training at all. I have therefore had the repeated experience of trying to get power fit from a relatively poor base. Instead of sprinting, I have always done this type of training by going up and down flights of stairs with a heavy backpack to simulate the demands of hoicking a sledge over an ice obstacle. With an expedition approaching, my training programme would start with finding the task impossible. After only one flight, certainly two, I would simply feel too sick to continue. My legs would turn to jelly, filled with lactic acid and painfully useless. But within a couple of weeks, and long before I became more muscular, I would find that I could run up and down the stairs again and again, perhaps repeating the flights twenty times with only thirty second breaks between them. It was always a pleasant surprise, and clearly due to improved motor recruitment rather than significant muscle growth or adaptation.

Even after full training, neither I nor the proper power athletes can recruit *all* the relevant motor units simultaneously. This is a safety mechanism since if all the units within a large, well-trained muscle were to fire at once, the tension could exceed the strength of tendons or even bones. The body has a built-in regulator to prevent such self-injury, although in circumstances of extreme psychological drive, it can be overcome. It is thought that this phenomenon may account for extra-ordinary tales of strength in which individuals have lifted cars to release loved-ones trapped beneath them in accidents. It may also be why the very best athletic performances do not occur during training but await great arenas and big events such as the Olympic Games. It is probably what happened when Bob Beamon broke the long jump record in the Mexico Olympics with a leap that was so much more than he had ever attained before or would ever achieve again.

Psychologically driven full recruitment can also cause less beneficial outcomes. During the 1995 'World's Strongest Man' event, there were a series of contests in which anaerobic muscle power was called upon in excess. During the arm wrestling, one of the finest of the competitors, fired by the pressures of the competition, generated such strong forces in his arm muscles that his humerus, the bone of his upper arm, sheared completely – a horrendous break in a bone much stronger than an average man's. All of us have muscles that are stronger than we know.

*

In addition to training, it is possible to influence sprinting or other power performance through modifications in diet. A great many individuals think that this means consuming large amounts of protein, as well as a host of other specialised substances, either legal or illegal in nature. Body builders are particularly gullible, gulping down handfuls of expensive supplements that are almost certainly entirely useless. If you eat a good mixed diet, your protein and micro-nutrient intake is probably way over your requirements even if you are an exercise fanatic, for the exercise itself will lead to a high dietary intake.

Until recently, the only things that really made a difference were anabolic steroids – illegal and the cause of violent personality disorders as well as liver cancer. In the last four or five years, however, research has identified one dietary supplement that can make a difference. Studies have shown that the short-lasting creatine phosphate stores of fast fibres can be made slightly larger in some people by consuming considerable amounts of creatine itself. In normal circumstances, your body obtains creatine from meat or fish or by manufacturing it. The final amount ending up in your muscles varies quite markedly, with people at the high-end of the normal spectrum having levels 50 per cent higher than those at the low end. If you have naturally high levels, eating additional creatine makes no difference to you, for it seems that there is a natural ceiling on how much your muscles can contain, but if you have naturally low levels, and you eat pure creatine in large quantities, your levels may rise to reach the natural ceiling, and with that rise may come an improvement in your ability to sustain a high work output. The amount of creatine is equivalent to eating about fifteen pounds of steak a day. Vegetarian sprinters or power athletes may particularly benefit from such supplements. They usually rely on their body's manufacturing capacity alone, and so are known to have naturally lower levels of muscle creatine. It probably explains why few successful sprinters have been vegetarians and, incidentally, supports the evidence from the design of our teeth and gut that evolution designed us as omnivores.

The improvement obtained by athletes using creatine or other dietary modification is small, perhaps just a few per cent, but it could be enough to make the difference between no medal and a gold. Nevertheless, nearly all top sprinters now use this legal supplement and it appears to have made little clear difference. Perhaps they all have muscles naturally high in creatine anyway, Hence, although world records do continue to be broken, the increases in speed over the last few decades are more likely to be due to better and more dedicated training than advances in nutritional thinking. A quite different situation exists with endurance sport. For endurance, we now know, nutrition can be everything.

FOUR

*

The Happiness of the Long Distance Runner

IT IS EARLY April in Greenwich Park and cherry trees laden with blossom provide a pink confetti blown by a stiff wind over a scene of mass madness. Beneath the trees thousands of people are gathering, not to attend a large spring wedding but to enter in one of the world's greatest sporting events. I, like so many others, am about to put my fitness to the test by taking part in the London Marathon.

It is not easy to run a marathon, even when one is well trained, and it has been only a couple of months since I gained a late entry. Such a short time in which to get fit is a far cry from that recommended. According to the running magazines, you should not contemplate starting such a race unless you are regularly covering ten or more miles a day, with a couple of longer runs sometime during the week. Due to lack of motivation and time lost to more pressing engagements, I have been doing less than half that. Nevertheless, throughout my adult life I have always kept up some sort of physical exercise – usually running once or twice a week – and I have ended up very fit indeed after some of my long Polar expeditions. I have persuaded myself that I am not starting from scratch but from a slightly trained base, fortunate enough to have been born with some degree of endurance fitness. My fibre mix must tend towards the slow side. If I do not reckon to go fast, I certainly want to run the whole thing without stopping. I have little confidence that I can do so, but I will do my best. I am nervous, but savour the excitement. As the time for the start approaches, the atmosphere becomes palpable.

It is not hard to see that most of the men and women around me would never make good sprinters. Although events like the London Marathon attract people of all shapes and sizes, none of those who look remotely serious about running could ever be mistaken for fast hundred-metre specialists. These bodies could not boast the body-builder musculature of the sprinter; instead they are men and women with the lean build of the oxygen burner. They are designed to go the distance.

As 9 a.m. approaches the contestants begin to strip down to their running kit and, in the bitter cold, cruise up and down to warm up. Watching them zig-zag through the throngs of people, I realise that they have run considerable distances before heading for the start queue. For me, such an enthusiastic warm-up would be foolish. The last thing I want is to add significantly to the daunting distance ahead. I will simply start slowly and warm up as I go along. I do some stretching exercises and jump up and down a bit. I hope it will be enough to stop tendons pulling. You should never run with muscles too taut from cold.

I need not have bothered. By now the start queue is at least a mile long – running from the gates of Greenwich Park back along the road that crosses Blackheath. Every hundred yards or so, there is a sign indicating an anticipated finish time. The idea is that you should be frank about your chances and join the line beside the time that corresponds to your ability. At the front are the real athletes – the men and women who can complete the twenty-six mile course in under two and a half hours. In the middle are those who have some chance of running the whole distance in four hours or less, and towards the back are those who will run, walk or even crawl to the finishing line some-how, come what may. Unfortunately, the length of the queue means that even if you are setting yourself only halfway back, as I am, it is going to take quite a few minutes from the firing of the starting pistol to crossing the starting line. By the time we are actually running, any warmth will be gone with the wind, and the benefits of stretches lost with it. The queue will also make any time round the course misleading – a disaster for those who want to set a personal best. For some of the runners who feel that their official time is important, the temptation to move up the line is simply too great. You have only to look at the front line group to realise that some of these contestants are in the crawling rather than winning category.

*

Even more than for sprinting, most humans seem to have evolved to cover long distances – maybe not running for mile upon mile but certainly at a walking pace for hour upon hour. The evolutionary need to develop such endurance is obvious. Without settled communities or agriculture, our distant nomadic ancestors went constantly in search of sparse food supplies or tracked animals that themselves roamed widely. No doubt most of them could have run the London Marathon with little difficulty – perhaps not at a modern competitive pace but at a steady lope which would take them through the distance in a perfectly respectable time. But if all our ancestors could perform such a feat, why is it beyond the wildest dream of most adults today? They see that tens of thousands of other men and women can complete a marathon, yet they seem to lack the self-confidence or motivation to do it themselves. Perhaps more significant, most people see no reason for wanting to do so.

We participate in sports from a variety of motives. Some yield rewards through the high levels of skill required, while others generate feelings of camaraderie from working in a team. Yet others give satisfaction through the human instinct to compete, but for most regular runners, none of these motives apply. There is little skill involved in running, even when competing at the top level, and a jog requires no skill at all. Although some men and women do enjoy the comradeship of running in groups, many are just as happy to go alone. Although some people aim for success in competition, most of those who do so have found their ability while running as individuals for pleasure. Overall, few runners are competing with anyone but themselves. So why do so many spend their time jogging round the roads, parks and hills throughout the year, and why are large numbers prepared to train for and suffer through a marathon? Studies of neuro-physiology suggests a possible answer – an answer with implications going way beyond marathons.

*

The starting gun fires exactly on time but, predictably, nothing happens. For some time my section of the line does no more than shuffle forward a few pointless yards, tightening the squash even further. Then, after several minutes trying not to tread on each other's heels, a ripple of action spreads towards us and our part of the gigantic human river begins to move. Slowly, the hesitant ripples become a steady current until, just at the point where we cross the start line itself, we finally break into a lope and move off with the tide. It has taken more than ten minutes since hearing the report of the pistol, but we are now on our way. Ahead lie the historic 26 miles, 385 yards to the finishing line.

*

In addition to electrical transmissions from nerve cell to nerve cell, the brain operates through chemical signals. There are huge numbers of these neuro-transmitters, and it has been shown that general states of mind such as pleasure, sadness, alertness and drowsiness can be dictated by their prevailing concentrations in different parts of the brain. Perhaps the best known example is the compound serotonin which seems to dictate changes in mood. Few of us can claim that we feel as happy through the long dark months of winter as we do in summer's brightness, and for some the winter sees a period of dark depression. This seasonal effect seems to be regulated by serotonin levels in the brain which are higher when we see a lot of the summer sun, and lower during winter's gloom. The changes are thought to relate to more rapid changes in another brain chemical, melatonin, which varies on a day to day basis as part of the careful programming of wakefulness and sleepiness designed to keep us in step with the daylight cycle. A circadian rhythm that evolved to wake us up and feel good when it is light, safe and our efforts can be productive, and to make us feel more drowsy and inactive with the onset of darkness.

As with all biological designs, natural selection would have led to near perfect control systems for these mood and alertness brain chemicals, but they were only perfect for the environment in which they evolved. Now they are unable to cope with the changes that have arrived since civilisation. The wakefulness regulator, triggered by the daylight cycle, still thinks we are hunter-gatherers, living in Africa, where winters are warm enough for us to be outside and the daylight cycle changes little with the seasons. It simply does not understand that many of us now dwell at latitudes where daylight hours vary greatly, and where the cold of winter tends to keep us indoors, even when there is daylight.

A bright day can see light levels outside approaching one hundred thousand candle power while normal indoor lights give out just a few hundred. A few hours of low winter's sun, passing through the windows of our buildings, is simply not enough for our serotonin system to realise that it is daytime and so, for many of us, winter sees a steady depletion of serotonin levels and a consequent lowering of mood. Even resorting to Prozac may not help, for it is only a second-rate means of doing nature's job. It raises serotonin levels to some extent, which is why it has some benefit, but it is nowhere near as effective as a daily dose of sunshine.

Running probably has no direct impact on serotonin levels other than getting you out in daylight. Even so, there are millions who can vouch for its beneficial effects on mood, which probably occur through the stimulation of other brain regulating chemicals, including the endorphins which are incompletely understood. Endorphins are widely spread throughout the brain and appear to grant pleasure and relaxation combined with a surprising clarity of thought. They also increase tolerance to pain and discomfort and may have evolved as part of our pain control mechanism. Clues to their action may be seen in the effects of extraneous substances that mimic them, and these include all the opiate drugs. Morphine and heroin act by artificially stimulating some of the endorphin receptors and so they too grant satisfaction, relaxation, and pain relief. The pleasures we derive from running may therefore be closely related to those sought by the serious drug-user or, to put it the correct way round, the pleasures of opiate use may be mimicking some of the natural pleasures of exercise, albeit more intensely.

Why running should influence endorphin levels is not clear, but a teleological approach suggests that if prolonged exercise was required for survival, and such exercise was painful, the body might well develop a pain reduction system triggered by exertion. Certainly all types of pain lead to elevated endorphin levels and running does produce constant low grade trauma. It is also possible that the increases are related in some way to the rhythm of exercise – either through the regular pattern of breathing or through the regular repetitive movements. All sorts of rhythmic processes can induce relaxation. Yoga is a good example, and either deep breathing or chanting mantras can also lead to a heightened sense of well-being. Music can have similar effects. Psychologists attempt to explain such phenomena by citing hidden memories of maternal heartbeat and unconscious feelings of pre-birth security, while physiologists demonstrate changes in cerebral blood flow with marked endorphin release. Whatever the cause, the root of the pleasures derived from rhythmic activities are complex, but there can be no doubt that the effects are real. Yoga, music, chanting, running – all can definitely promote happiness.

Talk of drug-like parallels suggests to some people that running must be bad for you. This is very unlikely since the unwanted side effects of drugs, especially addiction, probably occur because of inappropriate, poorly targeted and badly regulated dosing. Nevertheless, there is anecdotal evidence that exercise can become rather more than habit-forming, and many runners find that, once they start doing long distances, they cannot easily do without them. If they miss a day, they may become restless and dissatisfied. Certainly I have found this to be so when training hard for expeditions. Whenever a new departure has drawn near, I have gone from my one or two runs of a few miles a week up to long bouts of exercise on most days. Each time I have done so, I seem to cross a threshold beyond which something changes. Instead of having to force myself to go out and run again, I find it difficult not to and, once out, instead of constantly battling with a desire to foreshorten my run and turn for home, I find that I want to go on and on. Even ten miles becomes hopelessly inadequate, and I am only satisfied if I run for a couple of hours at a time. Although such drives might exert unwelcome social pressures, and I am sure there have been examples leading to divorce, there is no evidence that they are physically harmful. Indeed I believe that they may be of considerable benefit.

Once again, a teleological approach provides a possible explanation. If running genuinely tends to be addictive, the compulsion may serve some evolutionary need. Our ancestors had to exercise for much of their lives, and so natural selection is likely to have set them up to be in a good frame of mind while performing physical activity, while aiming to make them relatively soporific in between periods of exercise. The pleasure and freshness of thought that may be experienced during and immediately after exertion could therefore be another hand-down from our evolutionary development. It would certainly have helped our ancestors to become more effective hunter-gatherers.

Since exercise appears to induce a heightened state of thought, there are implications for those who do not exert themselves. It seems to me likely that, with no physical activity, you will inadvertently spend much of your life in a slightly depressed mental state – especially in the winter, when lack of exertion will be exacerbated by your lack of exposure to light. Runners may well benefit from a double whammy as they pound their way round on a winter's weekend. The commentators may pour scorn on them and suggest that exercise is unnecessary, laughable and probably downright harmful, but those critics are invariably talking of the physical side, claiming that the health benefits are unproven. Even in this respect, I would dispute their position, which merely reflects their need to justify inactivity and to support the delusion that they have no problems. But, in fact, this is not the main point. The real benefits of exercise go far beyond a toned up body and a bastion against heart disease. They go to the root of our thinking, and the sceptics simply know not what they miss.

*

For the first part of the race my body reacted in a familiar fashion – with pain and protest. Even though the crowding dictated a slow pace, my legs became tired and I was breathless. I had not completed a couple of miles before I was wondering whether I could keep going. Why was I doing it? Only past experience persuaded me that things would improve. If I stuck with the running, the breathlessness and fatigue would eventually go, washed away by alterations to my breathing and warmth in my muscles. It would not be long before my physiology made adjustments for my needs.

When you first start running, leg muscles are at only 37°C, or even a degree or so less on a cold winter's day. Yet it is at 38°C that they work best. Of course, once you start running, you generate a lot of spare heat and the muscle temperatures rise, but this can take a couple of miles, even on a warm day, and through that distance you will not feel it. The reason for muscles working better when warmer is that their enzymes are set up to operate at the 38°C level, which is an interesting evolutionary choice. If we had evolved to be sprinters in temperate climes, we would have been better off if our muscles worked best at 37°C or even slightly cooler so that they were ready for action at any time. The fact that it takes a considerable distance to bring them up to operating temperature supports the view that our origins were warmer and our usual needs were for endurance rather than speed.

The same sort of argument applies to breathing. When you start to run, your muscles need extra oxygen but your body is not set up to increase the supply immediately. For the first few minutes of a race, you develop an oxygen debt as you use more energy than aerobic systems can supply. It is only when oxygen in the blood has been depleted significantly and levels of carbon dioxide have risen that your brain senses these changes and sends instructions to set things straight. At that point, perhaps after a few hundred yards, you will begin to breathe harder and your heart will pump more strongly. But by then, besides having to meet the demands of your continued movement, you also have to repay the oxygen debt and clear the lactic acid that has accumulated. This takes some time, and so the first couple of miles of any run can be rough. It leads to an odd phenomenon. Most people feel less fatigued after running five or six miles than they do when they have run just one or two. Some inexperienced runners never realise this, and even quite reasonable athletes may believe that distance running is not for them. They have never run far enough to reach equilibrium and comfort and so have never found the capability that evolution bestowed upon nearly all of us.

*

As the discomfort slipped away I settled into truly relaxed running and knew that now I could carry on for many miles. I started to enjoy my surroundings. Moving along in this great throng granted me a special feeling of comradeship. The runners came from all walks of life, and were doing it for all sorts of reasons, but we all had a sense of sharing in something overwhelming that was mixed with those feelings of relaxation and clarity described earlier. The result was a state of mind that most people would not recognise. It was deeply pleasurable.

At the halfway point, however, things began to change. My legs began to weaken and complain once more, and I felt the return of breathlessness. The problems became steadily worse, and by the time I had been running for a couple of hours, and had completed around seventeen or eighteen miles, all pleasure had departed. My mind and body were hurting and my legs had become a strange mixture of rigidity and floppiness. In part, they felt like wooden sticks – sore, stiff, peg-legs. Simultaneously they were the useless, lax limbs of a rag doll. They had become wayward, quite beyond my control and, as I ran, one shoe or the other would brush against the opposite calf, causing deep red scratches to appear. The blood ran down my calves and vivid scarlet stains were added to my white running socks. I was barely able to go on running and kept tripping over my own feet. The tiredness felt similar to that which I had suffered near the beginning of the race, but I knew the cause was entirely different.

*

Fading at somewhere between fifteen and twenty miles is due to a shortage of glucose – the fuel that is favoured by muscles working hard. When sitting, lying or walking, fat is the preferred substrate, but when you start to run, things change. At an easy jogging pace, you are burning about half fat and half glucose, while at full pace you usually switch to glucose entirely. There are two reasons for this change in fuel usage. At low levels of effort, muscles can keep up with demands for ATP by burning the limited fat stores within them but, as you speed up, they need to mobilise fat from elsewhere, and the rate of transport in is limited. More important, however, the change towards glucose burning occurs because fat contains fewer oxygen atoms relative to carbon and hydrogen atoms than glucose, and so it takes more extraneous oxygen to burn it. This oxygen has to come from the air, and for the same rate of work, burning fat needs you to breathe fifteen per cent harder than burning glucose. If already working at levels that make you breathless, this addition is unwelcome and may well force you to slow down.

Most of the glucose needed for a marathon is stored as glycogen. The stores are mainly in the muscles themselves, although some glycogen in the liver helps to maintain blood glucose levels. At the start of a run, people usually have a total of 400 to 500 grammes of glycogen in their muscles which is enough to fuel them for about an hour and a half. After that the supply runs low. The liver then starts to manufacture more glucose, but only enough to maintain reasonable levels in the blood rather than to refuel muscles. As a result, the muscles run out and have no choice but to switch over to fat burning, despite the disadvantages. With the change, you inevitably feel more tired and breathless, and simultaneously, a lowish blood glucose can make you feel nauseated, sweaty and vague. This is an experience familiar to all marathon runners. You have hit the 'wall'.

*

I wasn't helped by the crowds around me. While I was still maintaining a rather unsteady run, the road became packed with people drifting along at a walk. Continuously picking my way round them, I was often forced to stop. It was like attempting to run along Oxford Street at Christmas time, and whenever I did stop, it took an enormous mental effort to break back into a jog. To me, these people were just selfish. Back at the start of the race they were the ones who had not been honest about their capabilities. I had started running at a pace that I thought would see me through the twenty-six miles in a little over three and a half hours – no great achievement, but then I am no great athlete. I stuck to that pace and, at the halfway stage, it even looked as if I might come in under that time. But then the crowd problem, which had started way back in the first couple of miles, became more and more troublesome. Obviously many of the competitors who had no realistic hope of running the race at even my speed started much further forward than I had. They didn't seem to care that, from the word go, they would be obstructing more realistic fellow runners, and by fifteen miles it was clear that many of them were not capable of running the distance at all. I could accept that starting well back in a queue, where it would take more than ten minutes to reach the start line, was not a good basis for an 'official' personal best, but time should not be what an event like this is about. Running a big marathon is about seeking a capability within oneself and sharing the experience with others.

As my legs became worse and my way more tortuous, I found myself mentally cursing my fellow participants, sneering at their pathetic athletic abilities. On later reflection, however, I saw that it was no more their fault that they were slower than my fault that I was not up with the leaders. How swiftly one can run a marathon is mainly down to the genetic hand of fate. At least they were out there, pushing their bodies to individual limits rather than accepting that marathons were only for the athletic elite. I pushed away my anger and turned my thoughts to the finish. It was now just six miles ahead.

Thousands would already have finished, some more than an hour ago. That was more than an hour and a half ahead of my likely finishing time. How did they do it? Despite some expertise in sport's physiology, I could scarcely credit it. I would find it difficult to run 800 metres at the speed they ran the entire race and many, who would not think of themselves as either elderly or infirm, would be unable to match their speed over one hundred metres. What mix of qualities allows men and women to run at the head of a world-class field?

*

The ability to run distance at speed is determined by both oxygen and carbon dioxide transport, although the two go together since they both depend on your circulation. The greatest amount of oxygen your body can take in, pump round, and then combine and burn with fuel is known as your maximal aerobic capacity and it varies with the size of your muscles, the number of their slow fibres, and the capacity of your lungs, heart and circulation. Aerobic capacities are very varied, and good endurance runners may have values of more than twice the normal, while the unfit may be limited to less than half the average. The values also tend to fall with age, although part of that decline is due to decreasing activity and is prevented if you continue to take exercise as the decades pass by.

Just as world-class sprinters may have inherited muscles containing 90 per cent fast fibres, world-class endurance runners may be even more genetically specialised, with muscles containing 95 per cent slow ones. As with sprinting, however, it is not just birthright that grants success. Leading, or even finishing the London Marathon has a lot to do with training, and a great deal to do with eating.

Endurance training improves the capacity for muscles to work without fatigue. It cannot alter the number of your muscle fibres, nor change fast ones into slow, but there are a number of intermediate types of fibre which can be made to become more like slow fibre types, although this may take many years. Along with some of the slow improvements to heart and lungs, it is this slow influence on the nature of intermediate fibres which allows good young 400 and 800 metre runners to become top class 1,500 or even 5,000 metre specialists later in their careers.

The main influence of endurance training on muscles, however, is not through changes to intermediate fibre types, but through improving the slow ones you already have. Just as sprint training makes fast fibres larger, endurance exercise increases the size of slow ones, and these may enlarge by as much as 20 per cent in just a few months. They then contain more fibrils, more glycogen, more fat and more enzymes, and in addition show improvements to their local oxygen delivery systems with an increase in the capillary blood vessels that surround them as well as a rise in the fibre content of myoglobin – a specialised oxygen-carrying molecule similar to haemoglobin in the blood. However, in contrast to sprinting, improvements to endurance performance through training are not so dependant on changes in the muscles alone. The improvements in local fuel supplies and oxygen delivery systems are useless if they are not accompanied by improvements in the shipment of oxygen and carbon dioxide to and from the muscles by the lungs and heart. Genes dictate the potential of lungs and heart in the same way as they do for muscles, but training brings phenomenal benefits.

The muscles employed in breathing are all very specialised, comprising almost 100 per cent *extra slow* fibres which have two or three times more capacity to use oxygen than any other muscle in the body. This is useful since they never need to turn to anaerobic systems which could lead to lactic acid build-up with consequent cramp and fatigue. Just a few fibres in the diaphragm have limited anaerobic capacity, and these can make themselves felt if used; they are the source of the familiar 'stitch'.

While training improves your leg muscles to carry you through the distance, your breathing muscles also become stronger, allowing you to breathe faster, with greater volumes and with less fatigue. The overall size of your lungs – mostly dependant on your body size – actually changes little, but maximum ventilation can increase from perhaps 120 litres of air per minute to more than 200 litres. This will obviously help you to run faster and further with less breathlessness. Yet, strangely, it is not usually deficiencies in your respiration that dictate whether or not you feel breathless. Indeed, it is not even a shortage of oxygen that makes your breathing laboured. That is due to the build-up of carbon dioxide from heavy work making the blood acidic. The acid then stimulates your breathing control to try to get rid of it. The extra breathing does help to clear this carbon dioxide but cannot solve the problem alone. You also need an increase in blood flow to carry the excess from the muscles to the lungs, and so a stronger heart is also required for distance running. Fortunately, the heart is the most trainable organ of all.

One of the most striking changes as your endurance fitness increases is a fall in your resting pulse rate. At first it may drop by one beat per minute for every week of training, and so in just a few weeks it can fall to well under 50 beats per minute. Further reductions, as low as 30 beats per minute, may then occur, although these take much longer. The reason for the slowing is an increase in heart size. Where an average resting heart might pump 60 millilitres of blood per beat, a trained one will pump twice that much, and the improved performance is even more marked during exercise when untrained hearts might reach 120 millilitres per beat but trained ones can exceed 200 millilitres. It means that, overall, the maximum output can go from perhaps 15 litres of blood per minute to more than 30 litres, and with the other improvements in muscle fibres and lungs, maximal aerobic capacity may rise by 20 per cent or more in just six months.

Maximal aerobic capacity, however, is not the be-all and end-all of endurance fitness. Equally important is your anaerobic threshold, or the level of exercise at which some of the work that you are doing needs to be met by anaerobic processes. If you are untrained, it usually occurs at about 50 to 60 per cent of your maximum capacity, and this means that when you work at levels higher than this for any length of time, lactic acid will accumulate. You will feel fatigued and breathless. In the well trained, on the other hand, the anaerobic threshold can become much higher, and lactate will not begin to accumulate until you reach perhaps 70 or even 80 per cent of even your improving aerobic capacity. It is a change that can make much more difference than anything else. Whereas much of your aerobic capacity is down to genes and is unlikely to improve by more than 20 to 25 per cent, the training change in anaerobic threshold from perhaps 50 to 80 per cent and will allow you to work very much harder. Indeed, it has often been exceptional anaerobic thresholds that have made for marathon winning talent. Alberto Salazar, the winner of many world-class marathons in his time, had a maximal aerobic capacity which was scarcely exceptional (70 ml/kg/min), yet he could work at 86 per cent of that without accumulating lactate. He ran the marathon in just two hours and eight minutes.

*

In addition to training, a diet to ensure good carbohydrate supplies is the key to success in marathon running. When you run, your muscles' glycogen stores are depleted, and upon finishing your efforts, mechanisms are switched on to refill them. These seem to become more efficient with repeated use, especially if they are given plenty of carbohydrate to work with. As a result, many endurance athletes have come to be thought of as living on pasta alone, although in reality they eat a wider variety of high carbohydrate foods.

In Britain, the United States and other western nations, most people consume a diet that provides about ten to fifteen per cent of their energy intake as protein, fifty per cent as carbohydrate and the remainder as fat. Since an average adult eats between 2,000 and 3,000 calories a day, this normal diet contains up to 1,500 calories of carbohydrate. An athlete in training often eats more calories overall and so, if sticking to a conventional pattern, might consume about 2,000 calories of carbohydrate. After a long run, however, they may have used up all of their body carbohydrate stores which, in a trained individual, could be more than 700 grammes, or 2,800 calories. A normal diet pattern would therefore be inadequate for their needs. Serious long distance athletes have little choice; they must eat abnormally large quantities of foods containing carbohydrates.

I use the term 'abnormal' with one important proviso. Although athletes consume large quantities of carbohydrate by western standards, their diets are actually quite normal in a worldwide context. What is more, they are almost certainly closer to the intakes of those ancestral hunter-gatherers who would have eaten primarily vegetarian fare. This is interesting from the teleological point of view. If our ancestors had no choice but to eat a high carbohydrate diet, and no choice but to undertake endurance exercise, it is hardly surprising that evolution set us up to achieve optimum endurance performance when on that type of intake. The 70 per cent carbohydrate diet used by marathoners to meet the exceptional needs of their sport is therefore quite natural. Without realising it, they have returned to the food patterns of our past.

Besides eating a lot of carbohydrate in general, endurance athletes also aim specifically to increase glycogen stores immediately after long events. They are able to do this because muscles which are nearly empty of glycogen have such an avid appetite for sugars that if flooded with plenty, they not only restock but actually overfill. This observation led to a dietary practice known as 'carbohydrate loading' which was used extensively in the 1970s and 1980s. At that time, endurance runners would deplete their muscles about a week before an important race by exercising hard over several days while deliberately avoiding all carbohydrate intake and then, with muscles completely empty, would switch to eating virtually nothing but carbohydrate for the remaining time before the event. With no exercise at all in that final pre-race period, their muscles were stuffed full for the start. Nowadays, such formal depletion and then loading with carbohydrate is less favoured, for it has been realised that if a high carbohydrate diet is eaten all the time, all that is needed to attain equally high glycogen levels is to wind down activity a few days before a race.

I did exactly this before my marathon, and during the race itself I tried to stave off the dreadful 'wall' by taking plenty of sports drinks as I went along. Such drinks are another modern accepted tactic to help runners through long distances. The rationale is that by taking drinks containing carbohydrate while on the move, stores will run down more slowly. It sounds logical but there are some problems. It is not possible to absorb carbohydrate as fast as the body uses it, and so you lose ground however much you consume. Furthermore, the carbohydrate in the drinks can have counter-productive effects. Because it leads to more sugar in the bloodstream, muscles are deceived into believing that there is plenty available and are switched towards burning glucose rather than fat. Simultaneously, the higher blood glucose also misinforms the liver about glucose needs and so switches off glucose manufacture. Finally, drinking carbohydrates to excess can make your gut feel pretty uncomfortable. All in all, the benefit from the carbohydrate in sports drinks is less than might be expected.

Even so, sports drinks can be an aid to marathon runners, especially by helping to prevent dehydration. Becoming dried out has a devastating effect on human performance, with as little as a two per cent deficit in body water leading to a ten per cent reduction in endurance capacity. Two per cent is less than two litres and this amount can easily be sweated in the first hour of a run. Over a long race, even when the weather is not hot, the body may lose many litres more. It is therefore vital that you drink as much as possible during the event to avoid this happening.

Even a commitment to drink plenty, and the provision of drinks along a marathon course, do not necessarily keep dehydration at bay. Our thirst mechanism is not one of evolution's greater successes, and we do not necessarily feel very thirsty even when fluid losses are significant. Furthermore, if you do drink a lot, the gut often cannot handle it because exercise has diverted blood away to the working muscles. If over enthusiastic, it is easy to end up feeling bloated, nauseated or in pain. Sports drinks are formulated to overcome these difficulties, but problems arise because they also aim to increase both fluid and carbohydrate absorption – targets that, unfortunately, are incompatible.

All sports drinks contain sugars and salts which are picked up and actively absorbed by the intestine. This helps to maintain sugars for energy and salts lost in sweat, and simultaneously pulls extra water through the wall of the gut by the process of osmosis. Drinking sports formulations can therefore help stave off dehydration for longer than drinking pure water, especially if they are cool and appealing in flavour to help overcome your poor thirst mechanism. Most of the commercial drinks, however, contain about ten per cent carbohydrate, a formula aimed more at the provision of energy than the prevention of dehydration. They are too strong to give best water absorption and, although it is difficult to be sure whether a shortage of glucose or becoming more dehydrated is going to be your greatest problem, I suggest that the latter is more likely to be critical. In my view, the average sports drink should be diluted two-fold.

*

The pain and discomfort continued as I staggered along the Thames Embankment beside the Tower of London. I had to keep gritting my teeth to prevent myself from grinding to a halt. My feet were sore, my legs had seized up and my brain felt fatigued beyond words. I felt sick, and guessed that I had a lowish blood sugar as well as low levels in my muscles. More than anything, I just wanted to stop and walk for a while, but I knew that if I did so I would lose something from the experience. I had entered the phase in which psychology was everything. Whatever I decided mentally, my body would obey. Although it was incredibly difficult to keep on running, the remaining distance was falling and finishing was becoming an ever increasing reality. I just had to keep going.

London's Thames bridges fell away behind me one by one, and as they did so, the distance remaining diminished steadily. Now, within six miles of home, I had entered realms with which I was familiar. Every step brought increasing certainty of success. This lifted my spirits enormously and the cheering crowds gave me more mental strength. The pain could not go on for much longer. I found myself chanting repeatedly an adage I had heard somewhere before: 'when the going gets tough, the tough get going'. It was a mantra that brought a revival in my running.

To be able to go on when it feels so uncomfortable must be the most essential quality for anybody who wishes to run such a race and, although easily belittled, I believe it is a quality with an import beyond running. As we approached Westminster, I and all the thousands with me had taken an opportunity to learn about, and even be impressed by, ourselves.

Past Big Ben, my pace was back up to the rate at which I had started, and on the road towards Buckingham Palace I became faster still. My legs were just as wooden, just as stiff, and just as painful, but we are designed to cope with such hardships when the goal is worthwhile. We can draw upon the very strengths that allowed our ancestors to survive.

The finish was fantastic. For the last two hundred yards I summoned up a blissful sprint for the line just to prove to myself I could do it. Then came the feeling of elation. The enormous satisfaction of achieving such things cannot be expressed, but as I lay down beyond the tape and gazed around at the thousands with me, I knew we all shared something special.

After half an hour or so, I left the Mall and hobbled to the tube. My legs were making slow progress when an old woman, also wearing running kit, drew up beside me. She smiled.

'How was it?' she asked, 'You look as though you felt it.'

She was right. I did, whereas she looked as if she had just come in from a stroll in the park. I grimaced as I made my reply before asking about her race.

'Fine,' she responded. 'Not as good as last year, but still under three hours.'

Three hours, I thought – so much faster yet so much older.

'Do you mind me asking,' I said, 'how old you are?'

Another smile.

'Sixty-eight,' she replied. 'Not bad for a pensioner.'

FIVE

★

Crossing Antarctica

BEYOND MARATHONS, ultra-marathons and other athletic events, the wild places of the Earth offer challenges to human endurance which can take men and women to their very limits. In mountains, jungles or deserts, a whole host of other factors are added to the problem of mere progress. They provide stern tests of the resilience granted by our evolutionary past.

The British have a long history of Polar exploration. Back in the glorious pioneering days of the early 1900s, British expeditions led the way in many attempts to unlock the secrets of Antarctica and, in particular, to become the first to reach the South Pole. This led to the most famous story of them all. In 1911, Captain Scott and four companions reached the Pole after toiling there on foot, dragging sledges behind them. Originally, they had planned to use motor tractors and Welsh mountain ponies, but the former had broken down almost immediately and the latter fared little better. Their companions had helped them by laying out a series of food and fuel depots for the return journey. However, at the Pole, the already weakened men met the cruellest blow of all. A Norwegian, Roald Amundsen, had beaten them to their goal, making an efficient journey with four companions, and with dog teams to pull their sledges. Mentally broken, Scott and his companions turned for home, but unlike Amundsen, who made a safe and rapid return, they were doomed. Two died early on the return journey, while the remaining three were to get more than halfway back to base before dying just eleven miles from comparative safety. They had almost reached the One Ton Depot when they perished in their tent as supplies ran out and a storm prevented them from travelling that last short distance to the replenishment they needed. They wrote down the whole harrowing tale right up to the moment of their deaths, and their bodies and diaries were then found the following year by a search party. They won their place in history for their heroism.

My first trip to Antarctica was made with British Antarctic Survey soon after I graduated as a doctor. I had always intended to use my qualification to take me on expeditions in addition to more usual practice, and in 1980 I spent nine months working on ice breakers and at one of the BAS scientific research bases on the Antarctic peninsula.

During that time, I met two men, Robert Swan and Roger Mear, who were going to change my life. Robert had gone down South with a mission: he planned to repeat Scott's journey – although without the dying part. In Antarctica he teamed up with Roger Mear and the plans changed. Roger, a professional mountaineer who had been involved in the move away from large Sherpa-assisted Himalayan ascents to small Alpine-style assaults, put it to Robert that the South Pole could be approached in the same vein. He saw a lightweight, self-sustained journey to the Pole with men pulling their own loads, but he believed that to return as well in such a mode was impossible. Instead, the walk should be a one way trip, to which Robert agreed. The pair of them, with a Canadian mountaineer, Gareth Wood, were to manhaul from the edge of the Ross Sea ice-shelf to the South Pole where they would be picked up by an expedition aircraft and returned to their start point. It was to be the longest unsupported Polar trip ever made by man.

At that time, the only way for a private expedition to get to Antarctica was by sea, yet the waters around Antarctica are frozen for nine months of the year. A ship can approach the continent only towards the end of the southern summer when the surrounding ice has broken out. For that reason, the expedition had to do the same as Scott's – to go in at the end of the summer and drop off the team who would set up winter quarters in which to live until the following spring. Because they would be so isolated, they felt it was essential to take a doctor with them. Failing to spot that I would be the only one without medical cover, I eagerly volunteered.

To help my career withstand this unconventional break, I decided to undertake some research while I was away in order to return with some benefit. My topic was to be the study of our body weight regulation through the winter and the physiological effects of the walk to the Pole. They were areas of research which, in the end, would influence my entire future career. Indeed, they were the areas that would lead to many of the themes in this book.

In November 1985, after months of winter darkness, the three men of the 'Footsteps of Scott' expedition set off, taking 70 days to reach the South Pole. My studies showed that the three men had each burned up between 5,500 and 6,000 calories per day. At more than twice the energy expenditure of an average adult, it represented hard work. Nevertheless, they completed the walk in good physical condition. It raised the possibility that one could do even more. Could Antarctica be crossed, from coast to coast, without the aid of other men, animals or machines?

*

Each of the 'Footsteps' men had lost around 6.5 kilogrammes (14 pounds) of body weight during their trek, and it was to these figures that I turned when considering the logistics of the much longer Antarctic crossing with Sir Ranulph Fiennes some six years later. How Ran and I met and evolved this Antarctic plan after several unsuccessful attempts together to cross the sea-ice to the North Pole is told in my book *Shadows on the Wasteland*.

Because of the much greater distance than the 'Footsteps' team had accomplished, Ran and I had to manage more hours of sledge hauling each day than they had, and the sledges would need to be heavier. Against the 'Footsteps' consumption of 5,000 calories per day, I therefore considered that our ideal intake should be as much as 6,500 calories each, but to haul a sledge filled with 100 days' rations containing that much energy was not a practical proposition. Instead, I decided that the best compromise was to eat only 5,500 calories and accept the loss of body weight. If it turned out that we made the crossing on an average expenditure of 6,000 calories each per day, we would lose no more than 6 kilogrammes of body weight by the time we reached roughly the halfway point at the Pole. If, on the other hand, the work rate rose to 6,500 calories each, our weight losses at the Pole would be doubled. Either way, it did not seem critical. A normal individual can lose up to 15 per cent of body weight without weakening much, and we intended to get fatter than usual before even starting. Beyond the Pole, with lighter sledges and heading downhill, we would perhaps suffer no further weight loss, especially if we could be carried along by the winds. These tend to blow away from the South Pole and so should help us on our way during the second half of the journey.

The idea of putting on some weight before departure did not go according to plan. For many months, Ran and I followed a policy of dedicated piggery – never missing a sweet, snack or lump of fatty food that came our way. The difficulty lay in our hard training, for it is almost impossible to put on weight while engaged in a lot of regular exercise. In fact, the opposite occurred and both Ran and I became lighter. This was a matter of some concern until we were saved by fate.

During our journey down to the Antarctic a combination of bad weather and a technical fault with our aircraft saw us trapped for ten days in Punta Arenas, a city in southern Chile. There we were unable to do anything but wait, go for the odd run and sample every restaurant in town. The result of this short period of gluttony was a gain for each of us of nearly 6 kilogrammes – a store that was later to prove very useful.

As well as limiting the calories we could take with us, the unsupported nature of our journey dictated the type of food we should eat. In order to gain maximum energy from minimum weight, fat had to be predominant in our diet since it provides 9 calories for every gramme compared to only 4 calories from carbohydrate or protein. Taking this to an extreme would have seen us eating one hundred per cent fat, with our ration of 5,500 calories per day weighing just 610 grammes. Pure fat, however, is not enough to sustain you. The question was: how much protein and carbohydrate were essential to our wellbeing?

Contrary to many people's belief, a lot of exercise does not necessarily lead to very high protein demands. A normal western diet already contains around twice as much protein as we actually require, and so our very high planned intakes would provide us with plenty even if the percentage of protein in the ration was quite low. How much carbohydrate we would need was a more difficult calculation. It would be beneficial to have high intakes for the same reason that it is helpful to ingest carbohydrate when running marathons – if working hard, the muscles prefer to turn to glucose. Although, like marathon runners, we would feel better if we ate a lot of carbohydrates, the benefits would be countered by their making our sledges heavier if they replaced too much of the fat in our rations. Furthermore, unlike marathon runners who may be working at levels of up to 80 per cent of their maximal aerobic capacity, and so feel terrible when they hit the 'wall' and are forced to turn from burning glucose to burning fat, I thought that we would plod along at levels of more like 40 per cent of our maximal aerobic capacity. At that work rate, muscles should be fairly happy to burn fat for fuel, so I decided that we could get away with a diet containing much less than the 70 per cent carbohydrate of the endurance athlete and opted for a ration containing 57 per cent fat, 35 per cent carbohydrate and 8 per cent protein. Of course, these were percentages of more than double the normal intake, and so the amounts of carbohydrate and protein actually consumed would be very large by normal standards.

Before starting our journey, we had hoped that the first two or three weeks crossing the 250-mile Weddell Sea ice-shelf would be relatively easy, since the surface would be flat and smooth. However, the Antarctic ice-shelves are made of glacial ice that runs down off the continent and floats on the sea, and the ice we encountered was split by deep crevasses. Pulling across such a fractured surface was incredibly difficult. With 100 days' food and fuel on our sledges, we each hauled loads of 220 kilogrammes (484 pounds), despite carrying a bare minimum of clothing and other survival gear. The friction from the ice was also far greater than we had expected. Even though our sledges were equipped with runners of the best non-stick materials, the slipperiness of the ice and snow was minimal. Even at low altitude and far from the Pole, the temperature was often as much as minus 30°C, which is too cold for downward pressure to cause the usual melting that provides a slippery mixture of ice and water under the runners. It was more like dragging huge loads over sand, and we struggled to make about one-and-a-half miles per hour. That meant pulling for between ten and twelve hours a day to make even reasonable progress. Twenty days passed on the ice-shelf before we met the edge of the land itself.

Once we reached land, we began the long climb up towards the Polar plateau, an ascent of more than 10,000 feet. The hard slog rapidly changed to bone-crushing toil. To our dismay, we found that we would haul our loads uphill all day, only to lose all of the height gained in just a few minutes of dropping into some valley that crossed our path. It was very disheartening. Adding to our misery, the winds blew constantly from the high plateau straight into our faces. We battled with white-outs that lasted for up to five days, seeing nothing and being pushed back by the force of the storms. The winds also carved huge dunes of snow and rock-hard ice ridges, known as sastrugi, around and over which we had to wind. As we climbed, it became steadily colder. Our pace grew slower and slower until we could no longer make even one mile per hour.

Mentally it was also difficult. At first I occupied the days of slog with thoughts about the outcome of the expedition, reminiscences of good times past, and all the things I would do when I got back. It was an ideal time for optimistic introspection. After a few weeks, however, I seemed to have thought of everything. I felt as if I had covered every subject so many times that I could think about them no more. Instead I became locked into the desperate tedium and hardship, and my mind was filled with thoughts of defeat.

There is no doubt that when one plans to take on another challenge of this nature, you cannot remember just how hard and painful the last one was. It is only after you have set off that the sheer hell of the work becomes real once more, and you remember the terrible discomfort last time. It is then that you recall why you said you would never do anything like it again, but by that time it is too late. It is not easy to announce to the world shortly after setting out – 'Sorry, I made a mistake.' You let down too many people, as well as your own ego, and so you find yourself stuck in the most difficult situation. In one sense you want to go on, but in another you are also desperate to stop. It leads to great temptation – to find a way out without losing face.

As a doctor, the obvious thing for me to do was to pretend to have an illness. Within a week of starting our crossing, all day had become occupied by my thinking of the best illness or injury to feign. For some reason, I thought that a sub-arachnoid haemorrhage – a kind of stroke – might be a good one, and I walked along wondering when exactly to put on the act. Each day I managed to resist the temptation to act it out there and then, usually making a decision to save it until we had put up the tent that evening. However, once inside the tent, tea and biscuit in hand and a few more miles under our belts, life always seemed rosy.

I am often asked what qualities are needed to make a person successful at such difficult undertakings. My answer is always the same and often surprises people. I firmly believe that one does not need to be particularly brave, strong, or even foolhardy; instead you need a very defective short-term memory. Of course, I am not unique in possessing such a quality, and indeed I would ascribe it as a general phenomenon in humanity. Which woman would voluntarily go through a second childbirth after the experience of a typical first? I believe that we are programmed to be incapable of remembering such pain, and that we cannot recall suffering to anywhere near the degree to which we can recall the times of pleasure. I am sure this is a quality that has been selected by evolution for a purpose. Optimism is needed for survival, and realism must be its worst enemy.

Lulled by my poor short-term recall, I knew that to give up without really needing to do so would mean spending the rest of my life with the knowledge of my failure. Determined not to damn my future, I broached the subject one evening.

'Ran,' I said hesitating. 'Do you ever think of feigning an illness to get out of this?'

He looked over, questioningly.

'No . . .' he said slowly. 'Do you?'

'Well . . . yes.' I was embarrassed. 'Pretty much all the time actually.

I keep thinking that I should pretend to have a stroke. So you should ignore me if I suddenly collapse.'

By telling him, I closed the door upon the option and so forced myself to continue the journey. I was proud of the admission, but there was a frosty silence.

'Are you shocked?' I ventured.

'No, not at all,' he replied. 'I may not have thought of putting anything on myself, but I do keep on hoping you'll be ill.'

*

In addition to fooling me into thinking I was enjoying myself, the relative comforts of the tent brought distraction through being busy. People often ask whether we took books to read upon the journey, but we did not: books are not edible. However, as well as their lack of calorific value, we would also have had no time to read them. At the end of each day's hauling, we had to get the tent up, cut snow blocks, erect a radio antenna, and get inside. We then had to de-ice all our clothing before we could light the stove on which we would make a hot drink, melt water for the next day's vacuum flasks and cook the evening meal. After eating the meal, we would make a radio call, do some navigation, write our diaries and attend to the dressings on our blistered, frostbitten feet. Often we also had to mend our clothing or take various medical samples for my research.

Although not all my scientific studies were to his liking, Ran undertook them with remarkable stoicism. The most frequent samples I needed were small nightly urine specimens which caused neither of us any problems. The little bottles also proved to be useful pieces with which to play the occasional game of chess or draughts. More problematic were the five times during the expedition when I had to obtain 24-hour urine samples and so collected all our output through the day and night into a large bottle. The volume was then measured before we took a small aliquot for later analysis back home. The purpose was to quantify what was being lost from the body over the whole period. These were important studies for my research but collecting the urine in a narrow-necked bottle, in a high wind at minus 50°C, proved to be a tricky business. We observed frost injuries never described previously! Yet as far as Ran was concerned, the difficulties with urine collections paled into insignificance beside the problems that blood sampling caused him. He really cannot tolerate the sight of a needle coming at him and will promptly feint if he is threatened in this way. Needless to say, he got around the problem by simply looking the other way, or by closing his eyes, while I took blood from him. The real test came when he had to take blood samples from me. He would not look then either.

As well as being comforting, conducive to memory loss and busy, the tent was also a place of friendship. It is my experience that to share hardship with another is to forge very strong bonds which will last a lifetime, and of course similar bonds are felt between members of a sports team on a Saturday afternoon. I believe that such camaraderie is also a result of evolution. Within the framework of conventional Darwinian 'survival of the fittest' thinking, it is not easy to see how altruism of this kind can fit. Like the inheritance of longevity, the selection of genes that help others to do well seems impossible, yet it can happen if it favours group success. A good analogy is a bird which gives a warning cry as danger approaches. Logic tells you that, if it stayed quiet as it flew off, it would be less likely to become the next meal and so hand down more of its genes. If the bird remained silent, however, other members of its species around and about, often closely related and so sharing its genes, would get no warning of the impending danger, and the predator would be more likely to take a life. The survival of the group as a whole would be more threatened. It is a process known as kin selection, and although rubbished by Richard Dawkins in his book *The Selfish Gene*, I still believe that it can operate.

Gene programming or not, it has been my experience that the more uncomfortable and dangerous a situation becomes, the more strongly camaraderie is felt. At the same time, I cannot claim that the hardships of our Antarctic journey were always faced with friendship. The pressures took a harsh toll upon our relations, and sometimes, while hauling, bitter disputes made things even more difficult to bear. Most of these quarrels stemmed from the fact that we did not go at the same pace, and often one of us would be left behind. These situations inevitably led to frustrations boiling over, though more usually they were resolved with humour. Seeing the funny side of a situation is another vital asset to survival, and I am sure that it is not just through chance that humour flourishes in times of adversity. Once again, I am confident that it is a witness to our evolution. The selection of the fittest has, since the time of our intellectual growth, always included the selection of the funniest.

*

On Day 67 of our journey we came close to disaster. The wind was blowing briskly when I stepped from the tent and the temperature had fallen below minus 40°C. The sun was still shining and visibility was good, but the weather was obviously changing. I reckoned it could turn really nasty. Earlier, when I had left the tent to answer nature's morning call, it had seemed quite comfortable. The air had been still and silent, and the sun's warmth had poured through my thin clothing so that there was no real need to hurry. Back in the tent, I had added only a cotton windproof and some fleece salopettes to the long underwear in which I had slept, but now, as I was packing up outside, I realised I had made a mistake and might run into trouble. Still, we were about to set off and the hard work should keep me warm.

Our morale was high. The South Pole was only twenty miles away and we were likely to reach it within two days. It helped to have it so close. I was in a better frame of mind than for many weeks past. Just to reach the Pole would be something – only the second time in history that it had been achieved unaided. Besides, we now knew we could go on. A complete crossing of Antarctica might still be far away, but beyond the Pole we would be setting a new world record. It was a satisfying thought and it made all our previous efforts seem worthwhile.

In retrospect, I can see that I should have stopped there and then to put on more protective clothing but, once out of the tent, the aim was to get moving as fast as possible. To stop work and start messing around with kit before even warming up was more likely to prove disastrous than helpful. The weather was deteriorating, but I pressed on, hoping that once we started hauling the heavy sledges, my efforts would rapidly dispel the chill from my limbs. That was how it had been on many previous icy mornings, some of which had begun in much worse weather than this. But this time, when I started moving, I felt much colder than I expected. My hands, in particular, with only gloves and thin mitts on, became increasingly numb. I curled my fingers into my palms in an attempt to restore the circulation but it felt for all the world as if I had been handed a bunch of frozen sausages. My fingers would be frostbitten if I didn't do something about it soon. I decided that I had to stop, despite the risk of getting generally colder, and go back to my sledge to get thicker outer mitts.

They were there on top of the load. Survival in the cold is an art, and anticipating problems is very much the key. But although the mitts were instantly available, I could not get them on my frozen fingers. I couldn't grasp with one hand the cuff of the other to pull it up. To make matters worse, although numb and dead to touch on the surface, my fingers felt unbearably painful inside. I found myself whimpering like an injured dog.

Windchill had now taken the effective temperature down to something nearer minus 80°C, although that was not unusual for the Antarctic plateau. Why had I not felt so cold before? One possibility seemed likely. The enormous efforts we had made were beginning to take their toll. We lost a lot of weight, despite eating the 5,500 calories a day. It must have added to our vulnerability and meant that our energy expenditure had been enormous.

Ran, who had started out behind me, came up to help, but even with his tugging on the cuffs, I still could not get my inert hands inside. As the minutes slipped by the situation became desperate. We were now both losing the function in our hands and getting rapidly colder. It prompted Ran to make an almost ultimate sacrifice. He took off his own still supple mitts and gave them to me, struggling to get his large hands into my stiff cold ones. Then he urged me to get moving again. It was a wonderful gesture.

Although my hands soon felt better, the stop had taken ten minutes and I was now deeply chilled. I wanted to step up the pace, to get warmer, but could scarcely move my sledge. My muscles had cooled to the point where they no longer functioned properly. Unable to move fast, I could not generate enough heat to match my further losses. I was getting colder and colder, and before long I realised that I had to stop again. I needed more insulation on my body as well in what had become a full Polar storm.

The only other garment I had was a fleece jacket. It would be tricky to get it on, especially as it needed to go under my windproof, before I became truly hypothermic. The dressing would also need some dexterity, which meant taking off my gloves again. Once I removed my gloves it was only moments before my fingers would not obey commands. This time it was the zips on the jackets that defeated them and failed to move. I was dismayed to see the tips of each digit turning chalky white and I could almost feel the ice crystals forming inside them. I began to panic. I was stuck, half-dressed, with literally freezing fingers when Ran caught up once more.

With his help, I managed to get my two jackets on and then to force my pitiful hands back inside the mitts, but the stop took another ten minutes, during which my body cooled further. I had reached a point too deep in a downward spiral, and as we set off again, my thinking began to fade. Although I kept on hauling for another half-hour or so, it was never fast enough to generate much heat. I steadily became colder still and more withdrawn until finally I began to wander from our course. Later Ran told me what happened next.

'When I caught up and called out, you didn't answer, although you passed the compass when I asked for it. Your head was covered and I couldn't see your face, but something struck me as odd. I asked if you were okay but got no reply, and when I pressed you again, you made a sort of unintelligible croak. It was then that it dawned on me you must be hypothermic.

'I tried to get you to help me put up the tent,' he told me, 'but you were just standing around doing nothing, so I put it up by myself and pushed you inside. I wanted you to light the stove but you just knelt there in the middle, completely still and unresponsive. You were in the same position when I came back a couple of minutes later after I had unpacked your sledge and thrown your sleeping bag inside. I shouted at you again to light the stove, and opened the box and gave you the matches. Then I went to get my things from the other sledge. I was another couple of minutes and, when I returned, you had started to light the stove and seemed a little more with it. Perhaps you were better for being out of the wind. Mind you, you weren't exactly compos mentis, and I had to get you into your sleeping bag and make you drink some tea. It was about half an hour before you started making real sense and to appreciate what had happened. When you started to shiver and complain of the cold, I knew you were better.'

In many ways, I was lucky to survive. If Ran had been out in front when I started running into difficulties, I would have been struggling with the gloves and the jacket alone. Inevitably I would have become more deeply hypothermic and probably wandered off and collapsed. Motionless on the Antarctic plateau, my survival time would have been a few minutes at most. By the time Ran realised that I wasn't behind him, his tracks back to mine would have been blown to oblivion.

On Day 68 of our expedition, the day after my hypothermic episode, we reached the Pole and equalled the current world record. We had completed nearly 900 miles of pulling and still had the means to go on, but it was clear that we were weakening and becoming increasingly vulnerable. We sat in the tent and discussed the situation. The Pole was the last place from which we could easily withdraw. Anywhere beyond this point might be difficult for a plane to reach. The fact that I was now so vulnerable to the cold raised a vital question. Was it sensible to go on or should we pull out of our expedition while it was safe?

Ran was seriously concerned. If I could become so weak before arriving at the halfway point of our journey, how would we fare later? I argued that my hypothermia could be attributed to a straightforward misjudgement; now that I realised I could no longer rely on hard work to keep me warm, I would wear more protective clothing whenever the wind put us at risk. Although Ran accepted it, the argument was somewhat flawed. By the time I did become ill, I was actually wearing all the warm clothing I possessed.

*

Many people are surprised to hear that on this walk across the coldest place on Earth we had so little insulation against the intense cold. In retrospect, I too am astonished, but it was not a complete lack of sense that had led us to take with us so little clothing. It is a fact that, while working hard, one can keep warm in almost any environment. At rest, a man produces about as much heat as a single 100 watt light bulb but the output can rise to that of a 2-kilowatt bar fire when working hard. It is truly central heating, and it is a common experience to wrap up well before a winter's walk and then find that, after only a few minutes, you are a sweating wreck; within a few hundred yards of the house or car-park you need to strip off all your spare clothing and carry it for the rest of the day. We didn't want to do that. While planning loads with 100 days of supplies in them, weight was the severest problem. The sledges must contain as much food and fuel as possible and, like books, spare clothes are neither edible nor flammable. We would cope with the cold by working hard or quickly getting inside our tent. It was a policy that worked well while we were of normal weight, but it began to go wrong as we lost more and more. In fairly thin clothing – the equivalent to normal office clothing in Britain – more than half of your insulation comes from the layer of fat beneath your skin. Ours was rapidly disappearing. At the same time, while we towed less, we could never stop moving for long. In each of our twelve-hour days, we stopped only twice for short five-minute breaks.

In any case, I am not convinced that thin garments were the only cause of my hypothermia. Although my symptoms were typical of severe cooling, my recovery in the tent was rather too fast for that. Perhaps I was also suffering from a low blood sugar – a condition known as hypoglycaemia – which can stop the temperature controller in the brain from doing its job properly. During our Antarctic journey, eating the high fat diet and working hard enough to consume our glucose stores, we had good reason to end up with low sugars. The possibility is also supported by some of the samples we took at ten-day intervals during the expedition and later brought back home for analysis. These yielded some extraordinary results, the most striking of which were abnormalities in our blood sugar levels. To put it simply, towards the end of the journey the levels were so low that we should have been dead.

Most of the sugar in the blood is in the form of glucose, with normal levels between 4 and 10 millimoles per litre. Above these figures lies the tendency towards diabetes, whereas below you are moving towards confusion, coma and death. Analysis of our samples showed that our blood glucose had been low from the very first day, in fact at the bottom of the normal range. During the last 30 days of the expedition they appeared to be impossible. On one occasion Ran had a level of 0.2 millimoles while I had one of just 0.3. In ordinary circumstances, these would be fatal, for it is thought that the brain cannot survive without a reasonable amount of glucose for fuel. Indeed, many experts with whom I discussed the results thought that they must be invalid and that we might have left the preservative out of the sample tubes. This seems unlikely. The tubes were loaded with preservative before our departure from Britain, and other results on the same samples were normal. It seems that we really did run these impossibly low levels and must have adapted in some way to cope with the situation.

One possibility would be for the brain to switch to using ketones for fuel in place of sugar. Ketones are chemicals associated with starvation, and it is known that they can be used by the central nervous system to a small degree. Perhaps we had shown that they can be used more extensively, and certainly such a capacity would have served mankind well through evolution. During droughts on the plains of Africa, or out on the Siberian steppes in winter, long searches for food would not always have been successful. The ability to keep on working in the face of prolonged starvation would have been a survival asset, and so perhaps we had merely demonstrated a capacity latent within us all. Yet, if we can all adapt and survive such low sugar levels, why is it common to see patients unconscious in hospital with levels nowhere near as low? The answer might lie in the speed of change. The hospital patients are often diabetics who have taken too much insulin or missed a meal, or patients who are both starving and sick. In such cases, their blood glucose levels will have dropped rapidly, and so their brains would have no time to adapt to an alternative.

*

The couple of weeks before reaching the Pole had seen our strength go into steep decline. This had to be due to our weight losses, although I was unclear how great these were. We could tell from the fit of our clothing that a lot had gone, yet it did not seem possible that we had lost much more than the top of my pre-expedition estimates of ten to twelve kilos. I certainly did not trust the readings on the tiny scales that we carried. All the weight measurements had been taken in the tent, on a floor that tended to be soft snow which allowed the scales to sink in when we stood on them. Still, my belief turned out to be wrong. At the American Scientific Research Station at the South Pole, we weighed ourselves on a piece of plywood board. Much to my dismay, when Ran and I undressed and stood upon the scales, I discovered that our previous readings had been accurate. We had each shed more than 20 kilogrammes (almost 50 pounds). It was a loss far worse than my predictions and way beyond the levels at which strength declines, a loss equivalent to a food deficit of around 120,000 calories. It meant that over the 68 days we had travelled so far, our average energy expenditure had been well over 7,000 calories a day. It did not bode well for our continued journey.

The scale of these weight losses led to more careful analysis of our energy expenditures after we came back to Britain. Our pre-packaged diet on the journey made it easy to obtain accurate estimates of daily energy intake, and from these and the measurements of our body weight losses we could calculate energy expenditure for different segments of the journey. The results were extraordinary. During the period between Day 20 and Day 30, for instance, when we climbed from the ice-shelf to the plateau, we burned up more than 11,000 calories every day. This was 5,500 calories more than we were eating – a deficit equivalent to total starvation while running a couple of marathons a day.

Our energy use far exceeded any measurements previously reported in the scientific literature. For example, levels measured on the Tour de France – adjudged to be one of the world's hardest endurance sporting events – showed that riders use only about 8,000 calories per day. It was difficult to believe that we had been working so much harder, but any doubts about our data were dispelled by a second means of checking the figures. We had also measured our energy expenditure by means of a sophisticated technique employing labelled isotopes of water.

Measuring energy expenditure in men and women as they move around and perform all the tasks of life has always been difficult, due chiefly to the fact that the equipment needed to make the measurements interferes with what the subjects in any study are doing. What is more, in many situations – certainly when crossing Antarctica – you simply cannot take the necessary measuring equipment with you. Recently, however, a technique has been developed for estimating how much carbon dioxide an individual is producing from the difference in the clearance from the body of two labelled isotopes. Subjects drink water that contains very high concentrations of heavy isotopes of hydrogen and oxygen. Fortunately, although difficult and expensive to come by, both are stable and so remain in the heavy form indefinitely without radioactive decay.

When drunk, the heavy hydrogen and oxygen become evenly distributed throughout the water in the body and then, as body water is lost, they also come out – appearing in the urine, in sweat, or on the breath. The heavy hydrogen is lost in this way alone, and so its disposal rate reflects the loss of body water. The heavy oxygen, on the other hand, disappears from the body in both water and carbon dioxide, and so tends to diminish faster. The difference in the disappearance rates of the two isotopes depends upon how much carbon dioxide the body is making. By measuring that difference following a dose of the two isotopes, it is possible to calculate carbon dioxide production which, as the final product of nearly all fuel burned by the body, can be used to estimate total energy expenditure. With no equipment at all, just samples of urine for later analysis, isotope-labelled water allows measurement of an individual's energy expenditure over many days or weeks.

We had performed the isotope study in two halves – taking one dose of isotopes on the day before we started and following their disappearance up to Day 50, and then taking a second dose for the latter part of the trip. During the first 50 days, the results from the isotope studies tallied almost exactly with those calculated from the energy balance data, the isotopes giving average measures of 8,500 calories per day for Ran and 6,700 for me, while the energy balance figures were 9,100 calories in Ran and 6,800 calories in me. Once we had seen that the method worked, it was also easy to take the data and work out the energy we were using up during each ten-day period. These results supported the startling figures from the energy balance data. For Days 20 to 30, when we made the ascent to the plateau, the isotopes gave daily energy expenditures of 10,670 calories in Ran and 11,650 in me. They confirmed the highest maintained energy expenditures ever documented – values that must lie close to the physiological limit.

When we looked at the data for the second part of the expedition, following the dose of isotopes on Day 50, the results looked dubious – not because they gave very high levels of work but because the isotopes that were meant to disappear from our bodies diminished until around Day 80, and then started to rise. Additional heavy hydrogen and oxygen was appearing from nowhere. At first we were at a loss to explain this phenomenon. Then it occurred to me that, after drinking the isotopes, some of them could have been transferred from water into other more complex molecules that were effectively locked away from metabolism. They were then re-released rapidly when, towards the end of the journey, our starvation led to us breaking up our body structure, the sequestered isotopes reappearing in the body water pool faster than we cleared them, so that levels rose rather than fell. Like the blood glucose levels, here was evidence that things were seriously deranged by the end of our journey.

*

Immediately after leaving the Pole we were back to drudgery and, in my case, back to the consideration of those feigned sub-arachnoid haemorrhages. Our weight loss made me wonder how long we could go on. Success looked very distant. There were still more than 200 miles of plateau to cross and then half as much again descending through the Trans-Antarctic mountains to the far side of the continent. Even after that our journey would not be over. The Pacific side of Antarctica is bound by a second ice-shelf, the Ross, and that meant another 400 miles to open water. Only then could we catch the ship that would be calling briefly in mid-February – little more than one month away. However, we now pinned our hopes on the wind to assist us.

The Antarctic is a high cold dome, with the South Pole up at 10,000 feet. Because cold air is denser than warm, winds tend to blow away from the Pole as the air sinks towards the coasts. On our way up to the Pole we had walked into these winds almost constantly; now, as we left it, we could use them to advantage. We carried with us sails adapted from parachutes. We would stand on our skis with the sledges behind us and the sails attached to our harnesses in front. As the wind filled the sails, we were off, rapidly building up high speeds. In theory, one could regulate this acceleration by opening up a central hole in the chute by means of a control line. But the chutes had been designed for wind speeds of up to thirty or forty knots, and down in the Antarctic the winds are often far greater. It had almost spelled disaster earlier in the journey.

We had not expected winds to blow in our favour until we were beyond the Pole, but shortly after we had started, they blew in the right direction. We decided to see if the sails could pull our fully laden sledges. With the wind speed rapidly building to about fifty or sixty knots, the answer was yes – very rapidly. We tried to slow ourselves down by making the central hole bigger, but that only made the chutes unstable. With no appreciable decline in speed, the chutes had begun to swing from side to side, obscuring the way ahead and catching on some local icy protrusion. Still at great speed, skier and sledge then went through the hole, creating a tangled disaster of strings, sledge and ski bindings. There was then the risk of frostbite as we tried to undo the mess with thinly gloved or bare hands.

Sailing on the wind, it was also difficult to discern hidden crevasses. On one occasion I had seen Ran, just ahead of me, fall to the left of a long open crack. Further to the left, the ground looked quite firm, and so I had steered that way and carried on. What I failed to realise was that he had not fallen over by chance. Hidden by a rise, the crevasse actually extended to either side of him, and he had thrown himself over to avoid going into it. He tried to warn me as I approached, but I was going far too fast, and both I and my sledge were pulled into the gaping chasm. I was very lucky to survive. Instead of falling freely, I was partially held up by my sail and, even more fortunately, I went into the crack where it turned a tight bend, just at the point where blowing snow accumulated to build a delicate arch of drift. Instead of plummeting into bottomless darkness, I dropped only twenty feet on to a snow bridge – winded but unhurt. The next moment I was terrified. Crevasse walls rose almost vertically above me and to either side, just a few feet away, blackness beckoned. The bridge was clearly frail, and it took considerable effort to quell my fear and to unload the sledge one ration, one fuel bottle, one piece of equipment at a time. Each of them had to be thrown up to Ran on the edge twenty feet above before finally the sledge was empty. Then, with the aid of a rope, I gratefully climbed out. It had been a narrow escape.

Soon after leaving the Pole, we were suffering even more malnourishment and became dreadfully debilitated. It was clear that we needed a boost if we were not to grind to a halt, and there was only one way to get one. We decided to consume an extra quarter ration pack every day. It took our intake up to over 7,000 calories. We would eat this much for however long it took to cross the remainder of the plateau, and then, when we reached the top of the glaciers on the far side, we would cut the food right back and try to survive on just half rations. It meant that as we went down and off the continent, we would be eating less than 3,000 calories per day. If the winds came soon, we might still finish the entire journey, including crossing the second ice-shelf; if not, we might at least succeed in crossing the continent itself.

Eating more, but still very thin, we struggled to maintain our core temperature. The idea that we could keep warm through our work alone now proved to be even more misguided. Day after windless day went by. It seemed so unfair. In the ten days after leaving the Pole, we managed to put less than 150 miles behind us. Eventually, on two brief occasions, we got some wind from the south and made good distances, although each time it became too strong for safety. We found ourselves falling and being dragged over rough ground at tremendous speeds. Our sledges turned over and were ripped along upside down. After becoming tangled in the rigging, we were numb with cold and my fingers became frostbitten, blistered and were weeping pus. In addition, I was hobbled by a fractured ankle. Ran too had problems with his feet. A graft from a previous frost injury had broken down early in the journey and the pain he had suffered ever since hurt him more than ever now. The area could well become infected and, short of antibiotics, it looked likely that we would be forced to stop.

Even so, the wind had moved us on. Early on Day 82, as the storm cleared, we could see the mountains ahead of us. It was a glorious sight that lifted our morale and we reached them two days later. Here our route began to slope. It was as if we had reached a gigantic weir and we followed the ice as it poured from the plateau into the huge Beardmore glacier. With the move into the mountains, the surfaces rapidly changed from wind-blown snow to a sea of gleaming ice-glass on which we frequently fell. Without the crampons that had seemed like a luxury we could not afford, we resorted to cutting up used fuel bottles to make grips, but the metal proved too soft. Grips made from stitched rope were little better, so we were forced to travel at a teetering run, moving faster than we would have otherwise dared.

As we descended the Beardmore Glacier, the gradients lessened and the ice became covered with soft snow once more. The sledges began dragging again, and we were back to a slog. With the extra work came more hunger. Half rations seemed pitifully inadequate. A few spoonfuls each and the pot was empty, leaving stomachs aching. During the day, we now ate nothing but two small chocolate bars and shared one thermos of soup. Often drifting through a haze of unreality, I was not surprised to learn later that our blood samples showed those dreadfully low glucose levels.

We became steadily weaker. Although the sledges now weighed less than 90 kilos, we could move them only slowly. Well over a thousand miles lay behind us, but our bodies had paid the price. Even the magnificent mountains could not lift us as we approached the last difficulty of our journey. Ahead, the glacier ran through a narrow channel between two high peaks. Choked with crevasses caused by the constriction, our path was blocked by an enormous hump-back of fragmented, disrupted and tortuous ice that extended as far as the eye could see. A diversion to avoid it would have taken days, so we had little choice but to take it on.

At first the slope rose steadily, broken by huge parallel cracks, but we found places where they were bridged and slowly we climbed. At length we reached a region of total chaos, where the ice was split in all directions and more of the area was black voids than white surface. It was terrifying. The sledges had minds of their own, slewing sideways, trying to drag us to destruction in gaping mouths. We made our way through like men in a maze, wandering back and forth, seeking bridges strong enough to bear our weight. It took four hours to move less than a mile, but finally we descended the far side. Ahead of us the huge ice river smoothed out as it wound through the last of the mountains towards the five-thousand-foot Gothic spired summit of Mt Kyffin. It marked the far side of the continent.

We were now close to completing our continental crossing and just days from our first big success. Only Ran's frostbitten foot could stop us, and it would have to be a raging infection to turn us from that goal. On Day 90, close on Mt Kyffin, we became determined to finish in one haul. We set off at 6 a.m. as usual but did not stop at 6 p.m. In the event, we kept on walking until two in the morning – passing through the 'Gate' at the foot of the Beardmore and off the Antarctic continent. I cannot describe how it felt as we set up camp on the Pacific ice-shelf for the first time. After three months of hell, we had achieved something special.

That night we discussed our situation. We had only eight food packs left, sixteen days of half rations if we could keep going on so little. If we could sail, we might manage to cover long distances, but we were far too weak to haul the sledges much more. It was difficult to know if we had any real chance of reaching the far side of the ice-shelf four hundred miles away, but neither of us could say it was impossible. Just four hours after our stop, we set out once more.

Progress was slow. The wind did not come and our legs were giving way. Along with our fat, we had lost much of our muscle, and our bodies were mere sticks with clothes hanging half empty. Whatever our determination, we could not win by manhauling; we needed the wind. Surely it would not forsake us.

It did. For five days the weather stayed bright, cold and still. The surfaces were smooth and stronger men would have made good distance. Yet, in ten hours – now all we could manage – we pulled our sledges no more than fifteen miles. We were crushed by the vastness of our journey, sapped by the weight of our loads, and our muscles were poisoned by the ice. On Day 95, with no wind at all, we gave up. We could do no more.

While Ran made the radio call for our pick-up, I went and stood outside. Our tent was pitched in the middle of a huge white plain and the sun was shining. To the south, a thin line ran back from where I stood to disappear beyond the horizon, towards mountains and wind-sluiced valleys. There it ran back up the glacier and then due south to the Pole. It continued on – straight for the rest of the plateau, and dropped tortuously through valleys, dunes and sastrugi to the ice-shelf on the far side and so to the Atlantic coast. It was the longest, unbroken track that any man had ever made.

*

Ten days later, we were back in a laboratory in England, facing a bank of tests that would give a fuller picture of the effects of our journey. Most of the studies were geared to look at two inter-related areas. How had our metabolism reacted to massive weight loss induced by exercise, and what had happened to our strength and fitness? The results demonstrated that it was more than just blood glucose levels that were pushed beyond accepted limits.

Despite our huge food intakes, Ran ended up losing 25 kilogrammes (55 pounds) in weight while I lost about 22 kilos (48 pounds). In order to find out how much of these losses were fat and how much were muscle, we measured our density – not in the intellectual sense, but as measured by Archimedes in his bath. Because fat is less dense than muscle, the fatter you are the more prone you are to float. Comparing your weight in air with your weight when immersed in water will therefore tell you how much of your body is fat and how much is lean. We performed these underwater weighings before and after the journey and discovered that our bodies had dropped from a fairly normal 16 to 17 per cent fat before starting to dangerously low levels of just two or three per cent after the journey. As the measurements after the walk were made when we had returned home and already eaten an enormous amount of food, it seemed likely that at the end of the journey itself we had been around zero per cent fat. It was surprising we were still alive, and little wonder that we had felt the cold so badly.

If you stop eating and just have water to drink, you usually die after about two months, even if you lie down and use as little energy as possible. This is known from famine victims and political hunger strikers in whom deaths usually occur when they have lost about one third of their body weight. It is also known that for the last few weeks of their lives, such starvation victims are unable to move much, due to sheer muscle weakness, and that at the time of death they still have some body fat. Why then were we able to continue, albeit slowly, when we had lost similar amounts of weight and had pushed our fat levels down to hitherto unreported extents?

Possibly the answer lies in the through-put of food in our unusual situation. Although our daily weight loss was often much more than would be seen in a person on a hunger strike, we still had a normal to high intake of protein rather than low levels or none at all. Our bodies therefore had enough to maintain essential organs, so that our hearts, livers and kidneys did not cease to function as they would in straight dietary malnutrition. Indeed, some of our measurements proved this directly. During our 24-hour urine collections, we had taken another isotope label – a heavy nitrogen atom attached to an amino acid. Amino acids are the building bricks of proteins, and by measuring the rate of reappearance of the heavy nitrogen in our urine, and knowing how much protein we were eating overall, it was possible to estimate the rate at which our bodies were building protein structures. Even as we starved, these levels were high, the opposite of what is seen if you are starving 'normally'. We had been able to maintain our functions in the face of weight losses that as a rule would be fatal.

The measurements of fitness and strength might have been expected to show that, with all our exercise, we became fitter. Instead our maximal aerobic capacity went down by about 20 per cent, and the strength of individual muscle groups dropped even further. This was noticeable. After I got back home, I could not walk upstairs without pulling on the bannisters, or lift my children, who were only very small. It was tempting to think that the weakness was due to the clearly visible loss of our muscle mass, but the causes turned out to be more extreme.

Before and after the expedition, we had specimens of muscle taken directly from the top of our legs with large cutting needles. These biopsy specimens were analysed to measure the activity of the enzymes in the muscles – the molecules that run the entire system within the cells. The analysis showed that, after the expedition, enzyme activity had declined by more than 50 per cent in both of us, whereas under training with normal nutrition they would be expected to rise by as much as 80 per cent. The internal structure and mechanisms of our muscles had literally been burned in order to survive.

Some of the other findings will be mentioned later in this book, but one other warrants mention here. The levels of our testosterone – the male sex hormone – declined progressively from the moment we started until the time we finished, and ended up so low that they were virtually undetectable. The explanation for this may be two-fold. It is recognised that testosterone levels fall with extreme physical stress, and certainly we could claim to have been through that. Testosterone levels, however, also fall in men who are isolated from women. In sub-mariners, for example, levels have been shown to fall steadily through their underwater voyages and return to normal just before they reach port. Not necessarily home, but any port – the anticipation of meeting any member of the fairer sex is enough to correct the abnormality. The effect that this drop in testosterone had on Ran and me is not clear, yet it may have been useful. Throughout our long trek, we never did fancy one another.

SIX

★

Marathon of the Sands

DARKNESS was falling as we came down from the high Atlas and the unheated and draughty bus became distinctly chilly. It had been cold all afternoon. Climbing up into the mountains south of Marakesh, we had left the warm sunshine of the Moroccan spring to move under wreaths of dark grey cloud. At the top of the pass we had stopped for some sweet black tea at a roadside stall. Strong winds had made it feel bitter and there had been a hint of sleet in the air, ice blowing off the persisting winter snowfields. Yet, despite the overcast skies and the cold, it was a spectacular journey, both beautiful and nerve-wracking. Through intermittent breaks in the cloud we glimpsed peak, cliff and distant plain, while our young local driver hurled his machine around the multiple hairpin bends, totally ignoring the precipitous drops. He saw them as just a part of his everyday world, while we could only experience the irrational fear that large drops exert – a magnetic attraction upon your vehicle, willing it to leave the road and plunge to oblivion. Each bend had brought an unwelcome reminder of our mortality.

Now, with the onset of darkness, we began to sense that the road was becoming less steep. Outcrops of rock that flashed by in the headlights became less obvious and instead the beams swung from tarmac across hints of desert sands. As the moon rose it illuminated an almost alien world. Around us, stretching away on all sides, were the dunes – their silvery, iridescent crescents echoing the source of illumination. Conversation that had lapsed during the descent from the mountains started up once more and the bus filled with the sounds of many languages. It also filled with an air of expectation. We had entered the Sahara at last and were now close to the test for which we had long been preparing – the Marathon of the Sands, one of the toughest foot races on Earth.

The bus drew into a sleeping village square that could have come straight from Spain or Italy, but as we stopped and the inhabitants emerged to greet us, it was clear that they were not of southern Europe. Their clothes, manner and complexion were of northern Africa. They were the people of the world's greatest desert.

The Tuareg are a proud race, the masters of the southern Moroccan Sahara – great travellers, story-tellers and survivors. Many are still nomadic, but others have chosen to settle, and their young children now swarmed up to the doors of the bus, anxious to see and shake the hands of these strange visitors. They also wanted to beg, not for money or food but for pens and pencils. It appeared that the local village school had no materials, and soon we had all given up our writing things. The children could have come from anywhere, though their elders were of striking appearance. All looked prematurely aged by their dwelling place, as if more than one generation separated them from their off-spring. Some appeared to hail from older centuries and, most strikingly, many had dyed their faces with pigment to a deep wizened blue.

After a long wait in the square, shivering in the desert darkness, a row of distant headlights could be seen approaching us across the sands. The vehicles were presumably on a track, though behind each one a huge plume of sand was rising. When the jeeps arrived, we transferred our belongings and packed into them. Engines revved as we headed out from the square on what would be an uncomfortable two-hour journey before we reached our start camp. The Tuareg children waved us away and a strange murmur arose from the blue-stained elders. They all had an air of deep mysteriousness, of knowledge beyond the imagination. They knew that the desert would test our strength, yet not in ways that they would ever seek.

*

Most people, when they hear about 'Le Marathon des Sables', are surprised that anybody should be so stupid as to attempt to run such a distance in the Sahara. Then, when they realise that it is not a mere marathon but a true ultra-distance race, they decide that the word 'stupid' is far too mild. Yet men and women can run much farther than is commonly thought, and they can do it in conditions of heat that many imagine would render us immobile.

The race is held annually, and each year it attracts increasing numbers of competitors. They come from all over the world, keen to pit their strength against one of the most difficult of environments, covering in one week the equivalent of five marathons – not on smooth London tarmac but over rocks, plains and shifting sand. To make matters more difficult, competitors must each carry a backpack containing all food for the entire week as well as a stove, sleeping bag, flares, torches, warm clothing and navigation equipment. On top of all that must go a supply of water although, in that case, it would be impossible to carry enough fluid for one day, let alone the entire week. Runners are therefore re-supplied at checkpoints on the route. At any one time, you only carry enough water for the next ten to fifteen miles. With a fairly heavy rucksack on the back, rough going underfoot, and the fierce desert sun, it is not a venture for the faint-hearted and many find it difficult to understand how such a race can be run. The answer lies in our remarkable ability to cope with heat.

Running through our bare skin is a network of blood vessels that dissipate heat through the processes of convection, conduction, and radiation, and if that is not enough, we also have tens of thousands of sweat glands which can automatically wet the skin to add the power of evaporation to our cooling. This ability to cope with heat stems from the millions of years of our early development in the hot cradle of Africa where our primate ancestors met a climate which would have cut down those who could not keep strong, fit, and above all fertile through the worst of the blistering summers. Natural selection melded early humans to make them extremely heat tolerant and when, 100,000 years ago, they started to spread to cooler parts of the globe, they took this potential with them. It has altered little since. We all possess the heat resistance of our early African forebears, and all modern races are equally capable of surviving, living, working, or even running across the hottest parts of the planet. We are each, after all, the naked ape.

★

We awoke early, the harsh rays from the east pouring into our wall-less Bedouin shelters and rapidly moving us from being cold to overheating as we lay inside. The stories of freezing desert nights had proved to be true, and our ultra-lightweight sleeping bags had been inadequate to prevent us becoming chilled. I had slept badly and now, at only 6 a.m., I was reluctant to rise until one of my companions opened his eyes and became wakeful enough to engage his brain. His sharp cry of warning brought us all to our feet. Around us, scuttling across the rugs, was a host of red and white scorpions. The thought of being joined in my bag by one of those barbed creatures was enough to cast off all remnants of drowsiness. Fortunately, the scorpions were as reluctant to be sociable as we were and, as if by magic, they vanished under the stones where they would spend the day avoiding the relentless sun.

We too would spend much of this first day in shelter, registering and preparing for the race. There were six of us from Britain – three were colleagues of mine whom I had persuaded to come and enjoy this mad challenge, and two others decided to enter the race when they heard that nobody from the United Kingdom had ever done so before. For me, the race provided an opportunity to get back to reasonable levels of activity and fitness, as well as another chance to visit an exciting and wild part of the planet. The heat would also make an interesting contrast to the conditions in which Ran Fiennes and I had crossed Antarctica.

The six of us were joining about one hundred and twenty other runners from a total of eighteen different nations, although French and Americans made up a considerable proportion of those we could see emerging from their shelters. It was our first proper view of the camp at which we had arrived late the previous night. Lines of Bedouin tents, each made of a large piece of dark cloth supported on wooden poles, filled about half the head of a broad stony valley that ran away to the south. The valley had a gentle slope up to the east and a vertical high escarpment of purplish banded rock to the west. The floor, of sharp-edged stones, was cut in places by deep, steep-walled wadis, presumably filled at times with water. At intervals down the valley stood clumps of palms or thorny bushes, their vibrant, rippling greens providing a striking relief from the otherwise earthen hues.

Each tent was carpeted with a threadbare coloured rug on which we had done our best to sleep through the chilly night, waking from time to time to rediscover that hips have no soft tissue cover at all. I had assumed that in the Sahara we would always sleep on soft sand, but it seemed that the Bedu shelters could only be well anchored on harder ground. The shelters were laid out in three sides of a square, with the fourth side made up of large administrative tents, one of which bore a big red cross. Glancing inside, I was reminded of a television portrayal of a MASH receiving station. Then we were summoned to a different larger tent by the aroma of baguettes, jam and coffee drifting across the camp. The organisers of the event were French and a continental breakfast started the day.

Nobody could enter the race without showing all of their safety equipment as well as their electro-cardiograms and medical certificates. Then their packs were weighed. Mine came in at a disappointingly heavy 12 kilogrammes (more than 26 pounds), and that was without any fluid on board. It was a weight somewhere in the middle of the range of other runners. Those who had taken part in the race before had particularly light packs, some achieving this by carrying no food at all except sugar and electrolyte powders to mix with their drinks.

Even more than in a single marathon, I had considered that eating and drinking on the move would be essential, as would eating well at the end of each day's race. My plan was to drink up to two litres per hour of rehydration drink, which contained some calories, and then to consume nearly 3,000 calories of carbohydrate across each evening and breakfast meals so that my muscle and liver stores of sugars would be re-stocked. As a result my pack weighed more than I wished, and encouraged by the voices of experience that were saying one could run on burning body fat alone, I decided I would cut back to a reasonable compromise. Less well trained than the true professionals, my body would not burn fat quite so happily, but I managed to reduce the weight of my pack down to 9.5 kilos (21 pounds) without any water.

Registration also provided us with guides to our route that would see us across the desert. Each day was divided into stages, with water at the end of each stage. The route appeared to be variably defined, for in places guidance was precise – 'follow the flagged tracks' or 'run along the edge of the wadi' – whereas in others there were more indefinite instructions. On Day 3, for instance, we were exhorted to

> . . . enter the band of dunes following a bearing of 176 degrees and maintain this for 18 kilometres. At the top of the final large dune, you will see the end of the stage on a bearing of 145 degrees.

It sounded to me as if this might be easier said than done. Following bearings across even rolling country is hard enough, let alone across steep sand dunes that all look similar and among which one cannot see far. Still, since I would never be up with the leaders, I also guessed that I would always have footprints to follow.

The total distance for the seven days was as expected, but daily distances were more variable than I had anticipated. Tomorrow, the first day, was to be a truly modest 15 miles, but Day 4 provided a mammoth 50 – almost a double marathon. A full single marathon was signalled on Day 6. Since, after the first couple of days, I would have run further than I had ever done in my life, I did begin to wonder if it would be possible for me to finish.

The heat added to my consternation. It became exceedingly warm, far worse than I had thought possible in April. By midday we were positively roasting, even lying shaded by the canopies. When the wind blew, it somehow made things worse, engulfing us in a wave of crushing heat rather than providing the relief one might have expected. Despite knowing I was well acclimatised, and that people had completed similar races before, it struck me that the climate might stop me before the distance did. When dusk fell and we settled down to sleep, these anxieties – added to the cold, the hard ground and fears of attentive scorpions – conspired to make it another restless night.

*

Unlike the earliest of our ancestors, and indeed the fish and reptiles of today, mammals and birds regulate their core temperatures very closely. This allows every cell to operate under stable conditions, using metabolic systems set up around enzymes that work best within narrow temperature limits. The human core temperature is set at around 37°C, although this varies with diurnal rhythms from about 36°C during the latter part of the night up to around 37.5°C in the late afternoon. These are also the resting values, and they rise considerably during exercise, reaching levels of 39°C or more if work is hard and the conditions hot. It seems that we are designed to cope with this for the optimal temperature for working muscle is well above the 37°C norm. Hence the value of the pre-race warm-up.

Theoretically, evolution could have led to any temperature at which to run our normal core level, with enzymes adjusted through eons to be at their most efficient at the temperature chosen. Yet nearly all warm-blooded creatures have similar resting core temperatures. Obviously, the higher a species sets its normal operating temperature, the more heat production is needed to maintain it, while the cooler the normal operating temperature, the more cooling capacity will be required. Perhaps our 37°C temperature reflects an evolutionary compromise with climate during the periods through which our ancestors lived. But if that were the only reason, one might have expected more variation in the core temperatures selected by different warm-blooded species that evolved around the world under very varied ambient conditions. It could be argued that the similarity is due to all warm-blooded creatures being genetically close enough to have had a common warm-blooded ancestor and that the basic enzyme blocks of life, and the temperature regulation system that goes with them, evolved very early.

There is another intriguing suggestion. The unit of energy used most widely in biology is the calorie – the energy required to heat up one gramme of water by 1°C – but its value actually varies according to the temperature of the water, even when insulation is total so that no extra heat is lost. The strict definition of the calorie is therefore the amount of energy needed to heat up one gramme of water from 15°C to 16°C, for above this temperature less energy is needed to make a one degree change. At least, less energy is required until you reach a nadir at around 35°C, beyond which the amount of energy needed to change water increases again. Regulation of core temperature close to this point allows changes in temperature to be made with minimum energy cost. Could this be the reason why evolution ended up dictating core temperatures of around this value?

*

By the time we reached the end of the first day my anxieties over failure were turning into certainties. Most of us were shattered. All of us had trained over greater distances than the 15 miles we had just run, but the rough ground had made it very hard and I felt worse than after the full London Marathon. Although we had not come across any really sandy areas, it had still been mostly soft underfoot, and where it wasn't, it had been viciously stony. Over soft ground, feet felt like lead and thigh muscles burned, while on the stones, toes were stubbed and it was impossible to attain any rhythm. Ankles also constantly rolled from side to side, threatening to go right over. I wasn't sure which was worse, sinking in or stumbling.

The start had been fun, with a carnival atmosphere overcoming our many apprehensions. We set off at 9 a.m., and although the sun could already be felt, it was quite pleasant being bathed in its rays. The air was fresh, with no wind, and the purple scarp of the west side of the valley could be seen heading away with almost undiminished clarity. About six miles on, it turned to the west, and there too we would turn before following the rocky wall down towards a plain. I heard various quips about rounding this huge bend on the inside line and of not getting boxed in at the start. The gun sounded and we were away, and with it our heat production rose. Unfortunately, the sun followed suit.

For the first hour I almost despaired of my pack. I tried to fix my mind upon the next segment of the run and to ignore both the weight and the pain in my neck and shoulders. It worked for a while, but the target I fixed on would come no closer. I soon felt miserable and uncomfortable once more. Around me were a very mixed bunch. Some were inter-national ultra-distance athletes, running hard for a purse of thousands of dollars and the title of champion of the 'Marathon of the Sands'. These quickly disappeared into the distance ahead. Others clearly reckoned to walk for much of the time, and I assumed that they planned to cover the distances only slowly in the hope that such self-control would permit them to complete each stage and push on through each day. They soon disappeared behind. The majority, however, were like me, average runners who wished to test themselves to their own modest limits. We planned to run for as much of each day as we were capable, without pushing so hard that we would end up pulling out. Of course, whether fast, slow or medium, each runner's aim was to complete the task, to exert their strength and will in order to finish all seven days.

Among the more bizarre runners were the two oldest competitors who had both run in all nine previous races. I ran for a while with Ibrahim, from Morocco, who was affectionately known as the Butcher of Fez. Although he spoke little English, we managed to exchange a few words on homes and family. At seventy, he was still active as a butcher, had twelve grown-up children and innumerable grandchildren. He was another prime example that age does not necessarily bring inactivity. The other elderly runner was one year younger and known as L'homme bleu. Every year that he had entered, he had adopted Tuareg dyes for the period of the race. Deep blue faced and grey bearded, he made a rare sight, and it was difficult to imagine him as his more normal self – a merchant banker from Paris.

After the first checkpoint at 7 miles things began to improve. Between there and the second checkpoint at 12 miles it was almost comfortable, and I was reminded once more of how the body will accept anything as reasonable for a while if asked to do so. After that, it was miserable once more, and the heat was beginning to tell. I grew hotter and hotter as midday approached and we headed out across an open plain, devoid of vegetation and of unknown size. Heat rising from the surface of dark stones made the landscape shimmer like rippled water. It seemed for all the world as if we were crossing a huge lake with the far side beyond the horizon.

I began to get cramps in my legs, and slowly I was forced to reduce my pace to a walk. Many others did the same, and we became a weary, ragged line following the trail of those who had preceded us. The sun beat down upon our heads and, despite wearing caps or hats, we began to suffer from headaches. The hats had cloth squares sewn on the backs of them to shade our necks, and we looked like members of the French Foreign Legion returning from a rout. Yet our enemy was the desert itself rather than some human foe. Eventually we all made it, entering the camp and feeling almost instantly better as our spirits were lifted by our first achievement. But how would we feel on the days following? We quickly found our water allowance and made for the shade. Tired, dry and overheating, it was time to drink plenty and eat what food we had. It wasn't all that long before we would have to start again.

*

To maintain our core temperature at 37°C, and certainly below 40°C which is dangerous, the heat we lose to our surroundings must be at least as much as heat we generate along with that which we absorb from the environment. This means that, when working hard, we must lose one or even two kilowatts through conduction, convection, radiation and sweating.

Conduction involves the direct transfer of heat through contact. Thus heat generated within the cells of the body is conducted outwards to the skin, passing through any intervening organs, muscles or fat. However, if conduction to the surface were the only means whereby spare heat could be removed, it would be both inefficient and uncontrollable. Deep inside the working muscles, the enormous heat production would make things far too hot. At the same time, far from working muscles, other parts of the body not in current use or too near to the surface might have too little heat to operate. What is needed, therefore, is a mechanism to transport heat from working areas where there is too much to other areas that need it, or to the skin to get rid of excess. This job is performed by the circulation, for as well as transporting oxygen, food and the chemical waste-products of meta-bolism, the heart, blood and skin act as a sophisticated version of the car water pump and radiator.

Once spare heat is carried to the surface, it is conducted away by whatever is in contact with our skin. Normally this is air and clothing, but sometimes we may be touching water, wood, or metal. Air is a poor conductor of heat, and so direct conductive losses tend to be quite small, especially if a layer of warm air is trapped against the skin by clothing. Water, on the other hand, is a good conductor, able to transfer heat about twenty-five times more quickly than air. The result is that when immersed in water, our heat losses are much greater than usual. A 25°C day in the garden is shorts, T-shirt and sweating weather, whereas 25°C in the swimming pool is only warm enough if you keep moving around. In fact, if you remain still in water at 30°C, you soon become hypothermic. Heat always moves from higher to lower temperatures. With water's good conduction, the 7°C difference between the swimming pool and your body will suck heat away at a rate exceeding the 100 watts your resting body will be producing.

Metals are even better conductors than water, and they can remove heat from our skin very speedily indeed. This is why metal feels so cool to the touch compared with wood. As long as the metal is at a lower temperature than your skin, it will conduct heat away from the area touching it and the nerves will tell the brain that the object is cold.

Conversely, when the metal is hotter than your skin, it will conduct heat in easily. A piece of metal lying in the sun, soaking up radiant heat, may literally be burning to touch, whereas wood in the same situation is tolerable. Conduction, as with all the heat transfer mechanisms, works both ways.

In reality, conduction rarely operates alone. Mediums such as air and water are fluid and so convection also comes in play. This is the removal of heat by the flow of one substance over another, and the air in contact with our skin almost always moves even when we are stationary. The movement occurs for many reasons. Even when we are lying still, the fact that warmed air is less dense means that air in contact with our bodies rises away from the skin to be replaced by cooler air. Most of the time we are not still anyway or, even if we are, the air around us is moving with the breeze. Convection and conduction take place constantly and in most environments we can easily lose more heat than we would wish. For that reason, we normally wear clothes to trap air against our skin and so limit conduction and convection to levels which are comfortable.

At rest in air of about 20°C, conduction and convection account for only ten to twenty per cent of our heat loss, with most of the rest taking place through radiation from every part of our body. Of course, we also receive incoming radiation from warm objects nearby and, more significantly, the sun. The sun transmits both light and infra-red heat. We absorb almost all the infra-red, whatever our skin colour or clothing, but the light energy is absorbed variably by different colours – light skin and clothing reflect most of it, while dark colours absorb it.

Even on the coldest of winter days the sun can provide enormous heating, and in places like the Sahara its effect can be immense. Worse, when out in the desert, the whole environment of air, sand and rock can all be at a higher temperature than your body surface and so will radiate heat into you. Indeed, once the environmental temperature exceeds 35°C, heat cannot be lost through convection, conduction or radiation and the input can be enormous. To run with a pack in such a climate might seem impossible since heat production will be well up on the bar-fire scale. How then do the competitors in the Marathon of the Sands survive? The answer lies in our heat loss through sweating, a mechanism that allows us to withstand extraordinary conditions.

*

After yet another uncomfortable night we were disturbed early, not by scorpions but by a sudden change in the weather. Half an hour before we were due to get up at 5.30 a.m., the air began to stir and a light breeze started to flap our shelters. Within minutes, the breeze rose to become a stiff wind and then a full-blown gale. It was an extraordinarily rapid change as the stars that had been gleaming with unearthly intensity disappeared behind the blowing sand. Indeed all was lost in an increasing, whirling stream of violence.

Under the shelters, the sand filled everything, and all but the heaviest of our belongings threatened to blow away. It led to another hurried exodus, after which there were pathetic attempts to eat without consuming too much of the streaming dust. There was then a long wait for the start at 8 a.m., but with a full sandstorm now raging, it kept being put back while the camp attendants wrestled with the Bedouin shelters. To get them down and packed up was difficult and not altogether successful. At least one was lost for I saw it – momentarily – whirling into the sky before it disappeared in a veil of dust. Even after the camp was dismantled the organisers were reluctant to allow us to set off, fearing that in such poor visibility runners could easily go astray.

We all stood around the vehicles, vainly seeking shelter from the painful blast of stinging crystals. Eventually it was decided that we should go ahead, although the route was changed to follow flagged desert tracks. It would extend the day's distance from 17 to 21 miles, but it would be much safer than running on a bearing when visibility was less than fifty yards.

At around 11.15 we departed, running down a trail directly into the storm. It was a bizarre experience. The sand was everywhere, and despite scarves around mouths and noses, we inhaled it, ate it, coughed it. Whatever we did with the rest of our clothing, the sand also got to our skin, and soon all moving parts became sore and chafed. It was desperately hard to run against the constant battering, and even when we changed direction the wind still seemed to blow against us. I had never run in such strong winds before, and reckoned that it doubled the work-load. It felt as though I was running steeply uphill, and nobody could sustain 21 miles of such work. There was one way out – we needed to help one another.

It was obvious really. Run behind another competitor, take turns to be out in front, and the load was cut tremendously. It was an example of slip-streaming taken to extremes and, even better, runners could form into long lines of single file, with only one front-man bearing the brunt of the gale. Every now and then that front runner could pull over and drop to the back of the line to enjoy some hard-earned rest. It worked brilliantly, and it brought a tremendous spirit of unity to the race – a feeling that was to remain for the rest of the event. The Marathon of the Sands is not so much a competition as an opportunity to challenge oneself. Everyone was in the same boat, and we knew it. If a lone runner was spotted through the gloom, the whole line would deviate from the course to go and collect him. Slowly each moving snake of runners grew longer, meandering across the desert and picking up new segments.

I also found it helpful to have no option but to stick with the pace. To drop out of the line would mean falling back completely, for it would be impossible to catch up again. There was therefore no choice of walking, even in the face of discomfort from a full bladder. It made progress remarkably steady until, after a couple of hours, the wind began to drop and the moving sands returned to the ground. This was of course welcome, but it now revealed a high sun and soon the stupefying heat of the previous day returned. With still more than half the distance to go, we came to a region of low scrub. It had trapped both today's and previous blowing sands and so had formed an area of small dunes. Although they were tiny by Saharan standards, we toiled through these little undulations as if carrying a hundred kilos, and with the hard exercise we cooked from within and without.

Wherever we looked, the harsh sun shone, casting black shadows on the rippled sand. Its heat was reflected from every surface, burning the face despite the dust that now coated every one of us. Although only the second day, our lips were already dry, cracked and swollen. I sipped at my drinks continually, my throat rasping from my laboured breathing. Slowly, as on the day before, I developed a thumping headache and I knew that my core temperature would be running higher than was safe. Once again, I also suffered cramps and nausea as I became increasingly dehydrated. My body was sweating at a rate that my drinking could not match, and it was with enormous relief that I finally saw the finishing flags fluttering up ahead.

*

Unlike convection, conduction and radiation, sweating can lose body heat at any environmental temperature. To evaporate, water requires an enormous amount of energy, energy that will raise the individual molecules from their relatively placid liquid activity state to the frenzied movement of molecules in a free gas. In comparison with the one calorie of heat that it takes to raise a gramme of water by one degree, the evaporation of a gramme of water takes 585 calories, and it is for this reason that wet skin is so wonderfully effective at cooling us. Most of us are familiar with such cooling when emerging from the sea or the swimming pool to stand in a breeze with wet skin. Whatever the heat of the day, you suddenly feel cool, or even cold, and remain so until the water runs out and once more your skin is dry. Then you have to return to reliance on your built-in wetting equipment – your sweat mechanism – to keep you cool, though not at the level you have just experienced. The sweat control mechanisms are set up to prevent too big a rise in core temperature rather than stopping any rise at all. We barely sweat at all until we are on the hot side of comfortable.

Although evaporation can extract an enormous amount of heat, it can only do so if the surrounding air is dry, at least, dry enough to have the capacity to take up further water vapour. Fortunately, the hotter the conditions, the more water the air can accommodate, and in such desert regions as the Sahara, the very hot dry air has an enormous capacity for vapour uptake. This is not true for all hot parts of the world, for in some regions it is hot and humid. If the temperature is over 35°C and the relative humidity is approaching 100 per cent, there are no possible routes by which to lose heat. Even if you do nothing you will inexorably heat up and eventually die. In such climates, man can only survive when nighttime conditions are cool enough for the heat gained during the day to be lost before the next one begins. Life is also made more miserable when sweat will not evaporate and instead runs down face and body in those dreadful salty rivulets. There is no doubt that humidity has far more effect on comfort than the temperature itself. You are more likely to succeed in completing multi-marathons in the ultra-dry Sahara at 40°C than you might were you to do the same in Britain at the height of summer.

Sweating can occur efficiently only if the body has enough fluid to spare, and in hot conditions, with high sweat rates, we will slowly dehydrate. As we do so, the rate at which we can sweat will fall and our tolerance to the heat goes down. In the Sahara, we were aware of this potential problem before we started. Some runners simply put the bottles of water supplied straight into their packs, stopping occasionally to fish them out and have a good swig. However, there was a better way. To maximise the amount of fluid that one can absorb, you need to be able to get at it constantly. Many of us therefore had drinking systems from which we could take small quantities even as we ran.

The system I used consisted of two 1.5-litre panniers built into the sides of my rucksack, with long flexible straws coming over each shoulder attached to the straps. The free ends of the straws were close to my mouth, with a bite valve to allow me to suck when I wished but remaining sealed at other times. As a result, I could take little sips often and I drank some fluid every minute or so throughout a day's run.

Unfortunately, even with such a system and a commitment to drink as much as possible, it is still difficult to get enough fluid in. Even when fluid losses are on a scale that will severely impair heat tolerance and performance, we do not necessarily feel thirsty enough to drink large volumes, and if we do, the gut finds it difficult to absorb. This is especially true when we are exercising and blood is diverted away from the intestine to working muscles and heat-losing skin. You can easily end up feeling bloated and sick. In fact, in circumstances such as those in the Sahara race, it is virtually impossible for the body to absorb as much as it can sweat.

Sports drinks can help to maximise the absorption rates of fluid from the gut. If they are cool and appealing in flavour, they also encourage you to drink plenty of them. They contain both dissolved sugars and salts which are picked up by the cells in the wall of the intestine. The absorption of these solutes then pulls water through the wall at a faster rate than usual due to the effect of osmosis. Such drinks can promote the intake of everything one needs – sugars for energy, salts to replace those lost in the sweat, and water to match your high sweat rates.

If you are going to use sports drinks during a run, it is important that you get used to them for it is easy to give yourself stomach cramps. My colleagues and I trained for some weeks before the Marathon of the Sands with our special backpacks, going for long runs with the panniers fully loaded. We were using a drink that came in a white powder form which we mixed with water to fill our packs. The habit led to some rather odd looks in the changing rooms from runners who felt that we must be bold drug-users. It also led to an amusing incident after I first received my specially adapted backpack. Running through some woods on an early training outing with my colleagues, who still did not have their drinking system, I stopped for a pee. While I emptied a very full bladder, back turned to my friends, I was also drinking from the backpack straw. Without thinking, I asked the others if they wanted a drink.

'No thanks,' came the reply from Richard. 'I have never had much liking for urine.'

★

The third day was to bring us into contact with full-scale Saharan dunes, and from the start at 8 a.m., we could see them lying in line across our path. After trying to run through the tiny dunes the day before, we knew that this would entail work in a different league, and we were already suffering from cumulative fatigue. Strange gaits also suggested that many of the competitors were carrying sore and blistered feet, and certainly my own were beginning to suffer badly. With the heat and the constant pounding, my feet had swelled up, and the shoes I wore were now tight. Several toe-nails were bruised and, indeed, were later to fall off. The soles were also blistering. It was not much fun until I had got moving, although once under way, it was remarkable how the pain disappeared. Evolution has set us up to ignore repetitive pain signals — a useful coping mechanism for the inevitable injuries of life.

As we ran, hobbled or walked through the first six miles of the day, the undulating obstacles came steadily closer and I began to be astonished by their apparent size. Although I had read of how large dunes could be, I nevertheless thought that it must be an illusion. Yet I was still quite a distance away when the lead runners reached them, and their tiny black-dot figures gave the dunes real scale. Some were hundreds of feet high.

At the checkpoint, I picked up four and a half litres of water which I added to more than one litre which I had left from the previous evening's allowance. Now I felt that I had plenty to cover the remaining distance, sand or no sand. I reckoned I could cross the 11 miles of dunes in about three hours, but on reaching the first dune, I began to recalculate. I had never seen anything like it, and certainly never experienced such a formidable obstacle to progress. Like all the others, it was a magnificent crescent of sand, probably a quarter of a mile wide and rising more than three hundred feet. It lay with the concavity towards me, and both the compass bearing and earlier runners' tracks took me straight towards the middle of it. I found myself entering a huge steep-walled bowl between the two curving arms of the crescent. Intending to maintain the shortest and straightest line, I continued right into the bowl and then attempted to scale the dune directly. Within a few steps, I realised it was hopeless. The slope was steep and so loose that I could only make progress if I tried to power up it using arms and legs simultaneously. But it was impossible to maintain the level of work needed to go on up and, as soon as I slowed, I found myself descending backwards on a fast down escalator. It was like skiing down through soft powder snow. Another approach was needed.

Glancing round, I noticed that the tracks I had followed dispersed to either right or left. Others had tried to scrabble at the face but their marks in the soft sand were nearly filled in. Since I was right in the middle of the bowl, it made little difference which direction I chose, so

I moved left, assuming that the best option must be to thread my way between the dunes. No such luck. The dunes lay in a carefully defined pattern to impede progress. The high points of each line seemed to be staggered so that the distances involved in skirting them were not feasible, and the sand in the gullies between was all very soft. The only way to get through was to ascend one or other arm of the dune crescent, making virtually no ground in the direction I wished to go, before dropping off the top of the dune at the point that would bring me down to the next crescent ridge. It entailed enormous additional distance.

This tactic for dealing with trans-dune travel only became evident when I reached the top of the first dune and looked ahead. There before me lay the desert of the movies – rank upon rank of these huge sand hills extending to the horizon. Some were small, some big, and some gigantic – far bigger than the one upon which I stood. And all were beautiful, having a tremendous purity of line and near perfect sculpted form. In among them I saw the figures of those in front of me, and from what I could see of their tracks, they appeared to be diverging like the branches of a tree. Not only could you not achieve a straight line, but at each dune ascent there was often a choice of 'best route'. Runners were making different choices and the race was rapidly spreading out. Of course, one might have thought that if everybody were following the same compass bearing, the net divergence would be small. But unlike the usual situation of travel on a bearing, when you can set your sights on something distant and then head for it, here there was no distant point on which to focus. The horizon looked the same in all directions, and although dunes varied in size, they all had the same shape and orientation. It was impossible to tell them apart with any certainty. As a result one could only follow the bearing by guesswork and intuition. Some competitors were clearly more intuitive than others.

At first I followed the ill-defined trails of some of the fifty or so runners ahead of me, but it only took a few branchings to discover that I was following what was probably a single set of tracks. Before much further, I disagreed with his or her choice of route and found myself climbing one of these grand dunes without any sign of previous passing. It brought about a strange pleasure, a sense of adventure, although when I descended into the quiet untouched bowl beyond, I also had strong feelings of apprehensive isolation. As I went on, the same anxiety was to recur whenever I was down deep within the sands, but every ten minutes or so I would reach the top of another dune and, if I waited a moment and watched, other figures could be seen emerging between the waves of sand.

A couple of hours passed and it became dreadfully hot. The sun was high enough to eradicate all areas of shade and the sand surfaces lapped up its radiation and then re-transmitted it. Within the still and silent bowls, temperatures were unbelievable. As I descended into them, I could feel the heat scorching my face like the blast from an open oven door. The sand was so hot that it was burning to touch, and this became a serious problem when going uphill, for I still had to use my hands on the steeper parts of some slopes. It was only on the dune summits that there was any relief from the baking conditions.

After a couple of hours I reached a very high dune and spent five minutes looking about. I had no idea how far I had come, nor how much longer I would need to go on. Even taking it easy with my water supply, I had drunk nearly three litres, so I would be hard pushed if the camp was more than two hours away. I could see quite a distance to the south – the rough direction in which we were heading – and even further east and west. Dots of people were everywhere, and I got the impression that many of the figures were standing watching others – as I was – probably trying to answer the same question. There was no information. I appeared to be about in the middle of the spread of competitors on an east/west axis, and was somewhat reassured. As far as distance to go was concerned, I could make no judgement – the dunes went on as far as I could see. If the journey took all afternoon and I arrived beyond the dunes at dusk, it would be hard to find the camp. Even more disconcerting, if it took that long, I would definitely run out of water.

I went on again, and a little later came down to a wider gap between the dunes that I hadn't noticed from my earlier viewpoint. It appeared to be a dry wadi that was running south-west. Although not precisely in the direction I wished to go, it would be much easier to make good speed that way. I could head back east once I exited the dunes. Out in the middle of what was essentially a small valley, I found several sets of fresh footprints and it was obvious that others had had the same idea. I followed them, feeling more optimistic, and even broke into a trot, but the heat was too hot to keep it up for long. The little valley began to wind and then turned further west. I would have to leave it to head south again. I was about to do so when I rounded a bend and came upon an extraordinary sight. Ahead of me, standing in the middle of a small flat area, was a single large tree. In it were two blue-winged rollers.

I was stunned. Since entering the dunes, I had seen no signs of life at all – neither plants nor animals. On previous days we had come across some scrubby bushes and the occasional palm, but this seemed to be some sort of conifer – perhaps a juniper, although I had never seen one so large. It seemed so odd to find it here in the heart of the most inhospitable area we had crossed. The birds were also the first I had seen other than a few scavenging vultures that had turned up at the camp last night.

In the shade of the tree there was other life. Five or six of my fellow runners were resting there, emptying sand from their shoes and redressing their blistered feet. I joined them and we conferred about directions, distances, and water supplies. None of them had any more ideas than I had. During the conversation I noticed something else. Many of the tree's roots were above the ground and, on one side of the tree, there was a deep hollow beneath them. Looking more closely, I realised that a small chamber had been deliberately excavated and in it were a couple of pots and a few bits of cloth and bedding. It was difficult to imagine who would shelter there, but some goat droppings around the tree suggested that some of the Tuareg must come here. Quite where they might be coming from and where they might be going was beyond me.

I set off again after ten minutes. The others had decided to wait through the hottest part of the day before they continued, but I was concerned that nightfall was the real threat. I walked on down the wadi, resisting any further temptation to run. Once back in the dunes, I would be struggling again, and I didn't want to overheat before that. As I suspected, it was only a few hundred yards before the valley turned even further west. I had to take the plunge and began to cross the dunes again.

I was alone in the sands once more. It was 2 p.m. and the sun was as strong as ever. I was down to my last litre of water and, even so, I had not been drinking enough. For the third day my head pounded, and now my vision swam. My mind seemed to wander from unrealistic pleasure in my situation to all the anxieties that could be imagined arising from it. I was not slowing down but crossing the dunes was so much further than I had expected. I felt a strange detachment from reality. Even when my water finally ran out, my intellect seemed to be a disconnected observer of events that were happening to my body. As I became drier and slower, this feeling of detachment became more and more marked.

It was 5 p.m. when I finally struggled to the top of what proved to be the last sand bastion. It was the highest dune of all, and from the summit I saw the huge hill fall away to reach a broad earthen plain. As

I had feared, the camp was not visible. I could not see far to either side for dunes, almost as large as the one on which I stood, stuck out farther on to the plain. Looking right and left as far as I could see and trying to gather where others were heading provided no help. Several figures could be seen as dots on other dunes to the east and west. That did seem to place me somewhere in the middle and suggested that the camp could not be too far.

Looking ahead again, I suddenly spotted that there was a track running parallel with the dunes about half a mile out on the plain. The camp would have to be close to this since the vehicles must have come in this way. Then I noticed figures moving slowly along the track in opposite directions. They would soon meet and pool their information, which might allow them a rational decision. Just then, a better clue presented itself straight from the story books. Over to the right, about a mile away, three large birds were circling. Even if nobody had died, the vultures were surely over the camp. I headed down to the track.

★

The regulation of core temperature in the heat is mainly sub-conscious and under the control of the cells in the base of the brain known as the hypothalamus. The cells receive information about body and environmental temperatures from specialised heat-sensitive nerves in both the skin and selected points deeper in the body, such as the walls of large blood vessels. These nerves are primarily responsive to temperature change rather than the temperature level, firing signals more rapidly when the temperature is climbing quickly. The sensation that the skin is hot is therefore much greater when it is changing between say $20°$ and $30°C$ than if it is static at $40°C$.

This is strikingly demonstrated by the experience of getting into newly drawn bath water, which almost invariably feels very hot, whatever you have done with the mixer tap. As you touch the surface – probably at around $40°C$ – the skin in contact with the water rises in temperature quickly, particularly on the toes, which tend to be cooler than any other parts of the body. The toe receptors undergo a huge temperature rise and fire rapidly however tentatively you step in. The brain receives impulses so fast that they are interpreted as: 'WARNING – very hot water'. In reality, of course, after a few moments of forbearance, the water feels perfect or even tepid. Then, as you gingerly venture further, each newly heated region repeats the warning message loudly, inhibiting faster progress.

The cells of the hypothalamus integrate temperature information from the body with their own temperature which is representative of the brain. If they detect that temperatures are on the high side of normal, they trigger physiological responses to increase the rate of heat loss and inform consciousness that it is getting uncomfortably warm. Appropriate behavioural decisions can then be made, such as removing clothing.

The physiological responses to heat include alterations in the blood flow to the skin and the triggering of the sweating mechanism. In the heat, nerves convey signals to open valves at the input end of the skin blood vessel network so that surface blood flow increases. This leads to a corresponding increase in heat that can be lost by convection, conduction and radiation. There are also increases in blood flow to deeper layers of muscle and fat which make them less insulating. More heat is then conducted from the core.

An overheating hypothalamus also triggers sweating via nerves running directly to each gland. In a non-acclimatised individual, the mechanism is activated when core temperature reaches about 38°C, or 1°C above normal. It is very effective as long as the air is not too humid, and with frequent exposure becomes more effective still.

Acclimatisation involves improvements in skin blood flow, the vessels open wider and sooner, but most of the benefit comes from increased sweating. On average, a non-acclimatised person sweats about half a litre per hour, and this can rise to as much as two litres or more in someone fully heat adapted. With the skin changes, heat losses may increase to nearly twenty times normal, resulting in a tremendous change in our capacity to live and work in hot conditions. It goes without saying that such sweating levels put one at risk of dehydration and, on this scale, losses can exceed the capacity for the gut to absorb water to replace them.

High sweat rates cannot be maintained for hours on end, not only due to fluid losses but because of the risk of salt depletion. Acclimatisation, however, does include changes that try to maintain salt balance. Sweat glands consist of groups of cells in the skin which can draw salt and water from the blood and pump it into the hollow centre of the gland from where a duct leads to the surface. The fluid formed by this process is at first extremely salty, similar to the blood itself, but as it passes up the duct, much of the salt is reabsorbed. When sweat emerges on the skin, the salt is much less concentrated. Still, problems can arise if an unacclimatised person is sweating freely. The speed at which sweat flows up the duct can be too fast for effective salt reabsorption and, as a result, when you first enter hot climates, you are at great risk of becoming salt depleted. After acclimatisation, however, the duct cells become far better at picking up salt, and even though the sweat flow may increase to four times normal, the cells still absorb most of the salt that is passing. The kidneys also improve their ability to hold on to salt so that, in the end, a fully acclimatised person in the heat needs no more salt than an unacclimatised individual in temperate conditions.

Acclimatisation can develop changes surprisingly swiftly, especially if exercise accompanies the hot climate, so raising core temperature from within as well as without. High levels of heat tolerance are acquired in just a week or ten days if core temperature rises by more than one degree for at least an hour during that time. Beyond that, further changes tend to be small, although there are some poorly understood long-term adaptations, taking years to develop, whereby the body becomes better at not over-sweating. It is interesting to find that all races have a similar capacity for both short and longer term adaptive changes, so emphasising the closeness of our genetic inheritance. There has simply been too little time for any of us to move far from our heat adapted African ancestors. After leaving a hot environment, the benefits of acclimatisation are lost progressively over just twenty to forty days. It is therefore entirely misleading to hear people speak of their heat tolerance as something that they 'got used to during the war'.

The key to managing the Sahara race was to become fully acclimatised before the event. All the competitors tried to do this, but I and my colleagues had the advantage of working in a research establishment with experimental chambers capable of generating any climate on earth. We ran on a treadmill in conditions similar to those we would meet during the desert race to come, and I topped this off by deliberately wearing more clothing than usual when I went for training runs and by having a hot bath each evening. In fact, the latter manoeuvre can achieve a considerable degree of heat adaptation on its own. It is a trick worth pursuing when going on holiday to a hot region if you want to be affected as little as possible from day one.

The very fact that we could all run in the Marathon of the Sands illustrates the effectiveness of human heat adaptation, but it is perhaps a less dramatic demonstration than that performed by an early Victorian scientist and showman. He used to stun his audience by entering an oven on stage at a temperature of something approaching 150°C. Scantily clad, he then remained inside for more than a quarter of an hour accompanied by a steak that went in with him raw. On his triumphant emergence, he would consume the steak, done to a turn, before his enthusiastic audience.

*

There was no avoiding it – over enthusiastic re-hydration from the night before was triggering urgent bladder signals and they just could not be ignored. Reluctantly I struggled out of my cocoon and winced as I stood up. My feet were sore and degrading badly. I limped over to the latrine. It was cold in the early morning light and there were beads of dew glinting on the outside of our shelter. I glanced at my watch and saw that it was nearly 5 a.m. It was still quite dark, and above the eastern horizon a star, perhaps a planet, was just visible on blue velvet. Beneath, the sky was already changing to a fierce orange glow. It was a harbinger of the heat to come.

Around the camp, other runners were struggling into wakefulness, many clearly suffering with their feet as much as I was. The miles we had already covered had taken their toll and some competitors looked pitiful as they hobbled around. Day 4 was to be the long one, the day that we had all dreaded since signing up for this crazy race. If we feared it before we left home, we feared it even more now. We were in poor shape to run two full marathons.

It had taken me more than seven hours to cross the dunes the day before and I was surprised to find, on finishing, that I remained well up in the field. It turned out that most competitors had run out of water, and many had been out long after dusk. From my sleeping bag, waking frequently to turn sore hips, I had watched anxious officials shining lights into the darkness, and occasionally a flash would be returned if someone was near. As they each approached, a ripple would go round the camp, and people got up to greet them. All the later runners, some of them arriving after midnight, had found themselves crossing a line to loud applause from their weary fellow competitors. I was astonished by this self-sacrificing tradition, but each arrival brought pleasure to everyone. Overcoming hardship together brings closeness, and it was accepted by all that these slower contestants, who had walked for much of the way, were having the toughest race. Those of us with some of the afternoon or evening to add to our night's recovery were lucky. These poor souls would have to be up again and making ready less than six hours after they had arrived.

The first stage of Day 4 was over a flat plain of baked earth and large rocks. I reached the first checkpoint at 9 miles after only an hour and a quarter, but the heat was already beginning to build. While less than confident that my body was back in proper fluid balance after the previous day's events, it was pleasant to find the second stage going so smoothly, and I was comfortable when I reached the second checkpoint at around 17 miles. From there I entered another broad belt of small dunes. While negotiating these – sometimes running, sometimes walking – I noticed the weather begin to change. A slight breeze increased swiftly and within minutes I was engulfed in another full-blown sandstorm. The field was already spread out and there were no companions nearby to form a mutually supporting snake as we had done two days before. My footsteps were blown into oblivion as soon as I made them. Fortunately, one of my team fellows, Mike Lean, emerged from the gloom and we kept close together, following the compass meticulously and hoping that we would spot the flags which extended on each side of all the checkpoints to lead contestants in. At last, a little over six hours into the day, we arrived at the 29-mile mark, checkpoint 3.

Here the route took a turn and, as we left, the wind blew more across our path rather than against us. Concerns about getting lost also diminished since we were now following a track which should take us through the whole of the next 11 miles. We were doing a slow jog, keeping fatigue at bay by moderating our pace. At one point the track passed a few buildings which comprised a ramshackle village. They were made of mud and straw and looked derelict. Although I saw no people, there were some goats and even one dog around. It seemed that we were in a generally more hospitable part of the desert. The way became scrubby, and through the blowing sand I began to see that we were passing trees of some size. Many of them were similar to the lone tree of the dunes, and occasionally one stood close to the track. Passing such a tree on the lee side produced an interesting effect. Since the wind had started and obliterated the sun, the heat had subsided and we felt quite comfortable. Out of the wind behind the trees, however, my skin temperature shot up, and although it only lasted a moment as I jogged past, it felt like jumping into a sauna. The air was not cool at all – it was simply dry enough, and the wind strong enough, to extract all the extra heat I was generating.

Halfway along the 11-mile track, Mike Lean, who was a better runner than I, disappeared ahead and I was left jogging slowly and alone through the still dust-filled afternoon. The track was vehicle wide, slightly sunken, and identified by the absence of scrub or trees. It took wild swings every now and then to avoid more established dunes, for the people of the desert have long accepted that you never fight with the sands. They simply move their roads, their fields and their buildings whenever the shifting desert dictates. One had to be attentive to spot where the track would go next. By mid-afternoon, my vigilance had lapsed.

I was trotting slowly in a daydream, following the vague outline of the route on automatic pilot. My mind was far from the desert, trying to banish the pain and fatigue as I made my way. It was a trick I had adopted in Antarctica, though there I had had less need to concentrate. Suddenly I came back to the present to find my way ahead blocked. In front of me stood a clump of scrub and trees covering the broad side of a large dune and, to my surprise, there was no sign of the track continuing. But I wasn't concerned: obviously the road went round the hill and either way I could just back-track and pick it up. I turned round, and rapidly turned back. The wind had built up and was suddenly blowing in my face. Even with sunglasses on, I found sand poured into my eyes. I knew from my compass that I should now be heading almost with the wind and so, if I cast to right or left along the side of the dune I had to pick up the route again. I decided to head right.

It was a struggle to walk in that direction, for the wind must have been up to something like forty knots and the whipping sand was painful as it scoured my legs and arms. I worked my way sideways, head bent against the blast, looking for the route. After ten minutes, I stopped. The dune was no longer in the way and, as far as I could see – not very far actually – the land in the direction that I should be going was flat. Seeing no track, however, I thought that I must have opted for the wrong side of the dune. I began to retrace my steps – or I would have done had they still been there. The wind had carried away all trace of my passing just moments before.

I was now walking with the wind on my left and the dune to my right back to where I started. Ten minutes beyond, I seemed to be well past the other end of the dune but still with no sign of the track. I began to grow anxious. I had only a couple of litres of water and could not afford to mess about like this. What should I do now? Should I stop and wait for the wind to drop and visibility to improve, or should I go on casting about, using up energy and water? I wasn't sure.

I would first go on a little bit further to the left, but just as I was going to restart my traverse, a flash of red caught my eye. About thirty yards away, back to the right, a runner had just disappeared behind a tree and, as I whipped round, he reappeared heading with the wind. He was either on route or very confident, and I raced off in pursuit. Within thirty seconds, I was on line behind him, and I was also clearly back on the track. I had been standing, pondering my best survival tactic just yards from where I should have been going. It brought home to me how easy it would be to get lost completely. I had been lucky, but another competitor was less fortunate.

We got the news of the lost Italian when we reached checkpoint 4. From there, the finish should have been another 10 miles but as I arrived, I noticed that the administration tent was surrounded by runners, either lying on the ground or standing, stretching and putting on extra clothing. Mike Lean was there along with another of my colleagues, René Nevola. René, the best runner of us all, must have reached here long ago. It made no sense, but he saw me coming.

'Too dangerous to go on,' he replied to my unspoken question. 'We're waiting here for transport.'

I sat down with him. Although the announcement was one of the most welcome things I had ever heard, I was very surprised. There were still a few hours until darkness, and in the race book it was clear that the route onwards continued to follow tracks. It was not like the organisers to let us off so lightly, but I for one was grateful and settled down to wait. As the afternoon wore on towards evening, it began to feel cold sitting in the wind. More competitors trickled in, and the checkpoint took on the appearance of an impromptu campsite, with runners huddled in sleeping bags. Towards dusk the rumours began to fly.

It appeared that a runner was missing. Someone had left checkpoint 3 but had not arrived here at checkpoint 4. He was a first class athlete, and the organisers were most concerned. Unknown to us all, they had been operating a safety system of notifying by radio the next checkpoint as runners departed the one before. In that way they knew the rough whereabouts of every runner on the course, as well as when he or she might be expected at the next checkpoint. Someone had failed to appear, and it was then that they had decided to stop the race.

Rumours began to harden into facts. The mystery runner was Mauro Prosperi, an Italian ex-Olympics Pentathlon champion. René, who was also of Italian parenthood and who spoke the language, had met him and this made it more real for him. You could see his concern. I too felt close to Mauro's situation since I believed I knew exactly what had happened. Nevertheless, although he was probably in for a rough time, I felt sure he would be all right. He might not have enough water, but he would be sensible and simply wait the storm out. All the runners carried flares and a signal mirror, and once the storm had settled, he would be found. The organisers even had a helicopter for just such an emergency. In any case, the race had been designed for safety. The course was out in the open desert but it was deliberately set to run parallel with roads that were never far away. We had been told just the day before that, if we became lost, we should just head south.

There was no possibility that we could miss the southern Moroccan east-west main trunk road.

*

We are designed to starve. Back in the evolutionary past, we could not rely on a constant food supply and so we put on fat in the good times and were able to conserve energy in bad. This propensity is unwelcome for those who wish to remain fashionably slim, but it is a useful adaptation for many peoples of the world and essential for the few that get stuck in a survival situation. Mauro could last for weeks without food, though not without water. Since the time when we were unicellular marine organisms, all processes of life have been geared to operate in a very specific set of liquid-based conditions, and the basic processes have not changed. Adapting to operate in drier circumstances, our evolutionary solution was to carry the right liquid – essentially the same as those primaeval seas – within our cells. To assist in this, we developed regulatory systems to maintain cell constancy, particularly using our kidneys to either excrete or hold on to water in order to match overall input and output, and to vary the loss of salts. Nevertheless, the kidneys cannot excrete no water at all. If you don't have about 700 millilitres a day to spare, you begin to accumulate the poisons of your own metabolism.

Getting lost in the desert is not a good idea, for there the daily water requirement is immense even if you are not running. The 700 mls needed for urine is combined with several litres needed for sweating to stop you overheating. Being acclimatised is only of value if you have plenty of liquid to spare. Mauro had left the last checkpoint with just three litres, and had presumably travelled and drunk part of this before becoming lost. He was therefore at no advantage over anybody plucked from the streets of London and flung into isolation in the southern Sahara without water. Of course his body would fight: it would reduce both urine and sweat rates to a minimum, but as soon as he began to become dehydrated he would risk becoming overheated, especially if he did not act carefully. He needed to combine his intelligence with his physiology in order to survive.

By and large, the human animal prefers behavioural means to limit heat exposure. When the hypothalamus detects a rise in body temperature, it informs consciousness and we begin to feel the discomfort we call getting hot. We then change our behaviour so as to limit heat production, decrease heat absorption from our surroundings and promote faster heat loss. When it is hot, we usually choose to do as little work as possible, and may even lie down. This minimises all muscle efforts, including those that normally work just to keep us upright, and this keeps the levels of heat production below even that of the 100 watt bulb. Lying down also maximises heat loss through convection and evaporation. Air passing over the lower body then becomes warm and humid and no longer rises past the upper body. Any advantage obtained from the horizontal position, however, may be offset if you are lying on hot ground or a mattress that insulates your back. Tropical dwellers found the solution years ago by inventing the hammock, which is a most efficient way of minimising heat production and maximising convective and conductive heat loss. It can even be hung in the shade, with the added touches of a fan and cooled drink.

Most behavioural adjustments to our heat balance are made through adjustments to clothing. Here a series of very complex interactions have some influence, and the choice of best clothing for the heat is not as straightforward as it might appear. You can dress up or dress down.

Dressing up involves protecting oneself from the hot environment through the use of plenty of material to cover the body, providing it with shade from the sun's radiant input, and insulating it from hot surroundings. This is the traditional choice of desert peoples, and so one might think that we would have been wise to follow their example by setting out on our Saharan run swathed from head to foot. Unfortunately, it is only the good way for people who plan to rest through the hottest part of the day, for the approach retains any self generated heat, severely limiting heat losses through sweating.

Dressing down means wearing as little as possible. As a result, you expose yourself to high heat input from the environment but allow sweat and the other heat loss mechanisms to work efficiently. It is fine as long as you have plenty of water, and effective in a modern desert run so long as you are not profligate. For traditional desert dwellers, who must minimise water use at all times, it would be useless. While he was lost without water, it was also inappropriate dress for Mauro Prosperi.

*

The fifth day was free for rest. We all needed to recover, and it would allow time for Mauro to be rescued. It dawned clear and still, and at 6 a.m. the desert quiet was shattered by the whine of helicopter turbines. We were confident he would be back within an hour or two.

People took a leisurely breakfast and after that strolled up to the top of some nearby high dunes to look out over the beauties of the Sahara without the pressure of having to move on. The day passed slowly, with just occasional excitement as the helicopter returned and everyone gathered to greet the lost one. Each time our hopes were raised, they were dashed again – it was only a return for refuelling. Finding Mauro was not as easy as had been thought. The jeeps were despatched to help search on the ground, with some going down to stake out the southern road. Wherever he was, he would surely hear the search helicopter and use his flares or signal mirror when it came near.

At about midday the organisers called for me. They knew I was involved in survival research and were clearly becoming concerned about how long it might take to find him. They wanted to know how long he might last in the desert, but so many factors were involved – the greatest of which was what he decided to do – that I found his chances unpredictable. If he was stupid and made great efforts to travel far and fast by day, his water losses would be immense, his heat stress appalling, and without fluid replacement he would collapse within a few hours. In a couple of days he would be dead. If he was wise, and waited in the shade, covering his body, he could last some days, although he would probably be unconscious after four or five, and then the chances of finding him would be remote. There was also another possibility. The fact that he had not been found on a clear day, lost from a route well marked and with a road running just south of it, was not a good sign. He might have pressed on to the point of collapse and then fallen, only to be buried in the moving sands. If that were the case, he would be dead already.

The question was academic and the organisers had no choice. They must search for at least a week, if not longer, and since they hadn't found him quickly, they needed assistance. The Moroccan authorities were notified, and the Army and Air Force moved in. The local tribes were also told, and many of them began to scour the country on foot. There were nearby, I understood, several small villages of the type that I had noticed the previous day. Unless Mauro had gone on to the point of collapse, there was no reasonable doubt that he would be found, or that he would find his own way to the road or a village. Everybody remained confident.

By evening it was becoming difficult to remain optimistic. The mood in the camp began to sink tangibly low, and by the time everyone tried to settle down for the night, anxiety was added to the many other reasons for sleeping badly. In the middle of the night, darkness was broken by the sounds of weeping. The Italian camp was going into mourning.

Next morning, Mauro had still not been found, but the race was to go on. Not all of us felt easy with that decision, but we all convinced ourselves that it was reasonable. There was nothing of practical use for us to do to help him. We could not all head off into the desert to look without the risk of others being lost. It was better to let the vehicles go on with the search, rather than tying them up with transporting us out of the sands. We would continue on foot towards our final destination. There was only a marathon on Day 6, and a 10-mile run on Day 7. After beating the long day, it felt as if little were left to do.

The reality was different. A marathon should never be taken lightly, and in the desert it is a very long way indeed. The fact that it was a familiar distance was also to have some surprising consequences. Nearly all of us had run marathons and so each man and woman knew roughly how long it should take them in cooler conditions on normal surfaces. That self-knowledge tended to set time targets which threatened self-control, and pushing hard could be dangerous especially as, for the sake of variation, the organisers had included a new physical test. That day, instead of dunes, our run would take us over a range of rocky mountains.

The race started well enough – 10 miles across a boulder-strewn plain, with an occasional wadi to break our rhythm. Yes, there was great pain as raw nerve endings screamed from blistered toes, heels and soles, but our brains were soon quite blind to their message for we had all made the conscious decision to ignore them. Even our general fatigue and stiffness dispersed as muscles warmed up, and soon we had all settled into a tiring but comfortable and familiar action.

As the rocky slopes of the mountains grew nearer, they looked more and more impenetrable. Huge cliffs of red and ochre sandstone merged with scarcely less steep, deeply-fissured slabs above. I wondered how we could climb, let alone run, over such an obstacle, but at the base of the rocks, beyond the first checkpoint and restocking with water, the route turned beneath the cliffs, rounded an outcrop, and entered a steep-walled canyon. It was a strange and wonderful place. Scarce rains from perhaps a million years had carved a way down from the mountains which now allowed us to go up into them. High walls with glowing bands of colour rose on either side, water-carved into Henry Moore sculptures. We followed the floor of the canyon, part sand and part rock, as it twisted and turned, scrambling over boulders that blocked our way. Inside the dark walls it was cool, and I was grateful for this as my body churned out more heat from the effort of ascent. But as I climbed, so did the sun. By the time the canyon widened and I emerged on the upper slopes of the mountains, it was beating down on me ferociously. The last thousand feet to the col were desperate.

I reached the col shortly before midday and found others who had got there before me crouched beneath a small rock overhang, trying to escape the fierce radiation while they drank some water. Few of us had much left. At the checkpoint before the canyon we had been given only three litres and, despite the climb taking less than an hour, our sweat rates had been immense. The ascent had cost me two litres of that water and I could expect no more until well beyond the mountains. It looked like a steep and difficult descent, and none of us could be wasteful.

The view from the top was spectacular. Behind was the undulating land we had covered in the last few days, while ahead was a flat plain with the next checkpoint at around 19 miles well visible. Far beyond, we could even see the finishing camp, shimmering through heat haze. But what really drew the eye lay further still. On the far side of the plain, a slash of emerald green divided it from a further range of arid mountains. Dotted within that shocking band of colour were the stark white buildings of Zahedan. It was the oasis of our finish – the point we would reach tomorrow with the final short leg of our run.

We started our descent, picking our way through boulders, slipping, stumbling, but grateful for the assistance of gravity even if it pummelled our sore knees. It was when we reached the easy terrain of the plain that many of us made a serious mistake. We could see the finish, a distant camp, pitched in the middle of an enormous shimmering lake of mirage. Knowing our marathon times from London, Boston, Paris or elsewhere, and paying too much attention to watches on our wrists, we began to run faster. The sun was up with all its power and it could not forgive this enthusiasm. As we ran across the last ten miles, even the further litre and a half of water provided at the last checkpoint could not prevent our problems. Bodies were churning out waste heat so fast that we simply could not shed it all. Simultaneously the ground, the air and the sun poured more heat into us. Hyperthermia became inevitable.

For me, the sensation was strange but familiar. As I made ever slower progress towards the finish, I found my thinking becoming detached once more from the discomforts of my body. Yet with this detachment there came a deep sense of unease. By the time I crossed the line I was dazed, nauseous, and so very restless that I was unable to sit down in the shade. I ended up pacing around like a man with an insoluble problem. I felt as if I were melting from both within and without and might soon be reduced to a pot of tallow. All the same, feeling hot was a good sign that my brain was still working. It understood the danger that I had put myself in and it was doing its best to let me know how stupid I had been. When seriously overheated, this is not always the case, and in others things were going seriously wrong.

My colleague Richard, the man with no taste for urine, was a younger and better athlete than I, but he turned out to be particularly vulnerable to the heat. He had suffered terribly on many of the earlier days and now lay well behind me in the rankings. This hurt his pride and so he had been determined to run a good time in this marathon. He did, and entered the camp way ahead of me. Yet as he crossed that final unforgiving plain, he ignored the messages from his own body even more than the rest of us. Rather than slow down, he maintained a very fast pace as his core temperature soared. Eventually it reached the point at which the control systems began to fail, and here he became exposed to an extraordinary design fault in our evolution – a fault that has caused many deaths in the past.

*

When Richard entered the camp he was close to collapse. Recognising that something was wrong, officials took him straight to the medical tent. There he joined others suffering from a wide variety of problems; some were hot and dehydrated, while many suffered from painful feet, muscles or tendons. Many of the heat victims appeared to be in a worse state than Richard – sweating, flushed all over, and complaining vigorously while calling for water. Richard said little. If anything he felt and looked cool, and it was easy for the young French medics to ignore him. Although they had all attended a special course to teach them about the sort of problems they might meet in this unusual race, the course had been brief and held two months back. Now they had forgotten a small but important point. When people overheat they usually suffer from a condition known as heat exhaustion, which is largely due to salt and fluid depletion. They feel sick and weak, but continue to complain of the heat, and when examined they are flushed and sweaty with the fast pulse and low blood pressure of dehydration. But, if left untreated and they become hotter still, things can change. They go on to get heat stroke and behave quite differently.

With heat stroke the body temperature rises so far that the control systems become wayward and inform the brain that the body is too cold instead of too hot. As they do so, not only do attempts to keep the body cool cease but the mechanisms designed to maintain body heat are activated. Victims suddenly feel cold, shiver and their skin becomes pale and dry. These are changes that can spell disaster for a patient in an intensive care unit, let alone for a man who has just

1, 2. At 72, Helen Klein undertook the first Eco-Challenge – a 300-mile multi-sport adventure race designed to test accepted physical limits with wading and swimming through miles of flooded canyons as well as horse-riding, running, white-water rafting, rock climbing and canoeing.

3, 4. *Left:* Differences in muscularity – Linford Christie, a typical anaerobic power athlete, and Liz McColgan, a lean oxygen burning endurance runner.

5. *Below left:* Thousands of ordinary people participating in the London Marathon testify to the endurance capacity within us all.

6. *Right:* Sir Ranulph Fiennes during the crossing of Antarctica – a 95-day expedition on foot which tested the limits of human endurance.

7. *Below:* The journey came close to a premature end when Mike Stroud and his sledge went into a crevasse on Day 3.

8. Approaching camp during the 130-mile Marathon of the Sands. The ability to run in the Sahara heat illustrates our extraordinary physiological capacity to cope.

9. Chaos in the medical tent at the end of a day running in the Sahara.

10. We have far less physiological capability to cope with the cold, but can use clothing and heat generation from hard work to protect ourselves.

11. Mike Stroud with Emperor penguins in the Antarctic winter darkness.

12, 13. We share 98.4 per cent of our genes with the chimpanzee and have bodies designed for regular exercise and a largely vegetarian diet, but modern lifestyles can lead to obesity and heart disease.

14. The British Columbia Eco-Challenge team – Ranulph Fiennes, David Smith, Rebecca Stephens, Vic Stroud, Mike Stroud.

15. Crossing a river using a Tyrolean traverse.

16. *Below:* Mike Stroud with his 70-year-old father at the end of their race. It proved that age need not be a bar to physically demanding pursuits.

17. Running by the Magellan Straits on the first of the seven global marathons.

18. Celebration – the finish of the seventh run in New York.

run a marathon across the Sahara and is lying in the corner of a tent.

If the victim has progressed to heat stroke through the more usual heat exhaustion, there will be strong clues to their condition, but sometimes it is possible to get heat stroke without going through the heat exhaustion stage. If you heat up quickly enough, or drink adequately while under great heat stress, the controller fails while you are still quite reasonably hydrated, and this is exactly what had happened to Richard. His core temperature had probably been quite reasonable until he reached the plain and he had felt fine, maintaining a fair fluid intake. Then, when he stepped up the pace, his core temperature rose swiftly. Entering the camp less than an hour later, his temperature control had failed, before very much fluid had been lost. By the time a doctor examined him, he had also been encouraged to drink and was marginally rehydrated. He was not thirsty, did not complain of being too hot, did not appear to be dehydrated or sweating and had a relatively normal pulse and blood pressure. Certainly he looked unwell and appeared pale and vague, but nevertheless he seemed much better than many around him. The doctor told him to lie down and said she would attend to his sore feet later.

Richard was critically ill. Over the next couple of hours his temperature control continued to malfunction and slowly he became worse. He began to feel increasingly cold and his automatic anti-cooling systems became more active. They cut the blood supply to his skin still further, and the command went out for his muscles to start heat production. Noticing that he was lying covered in goose pimples and shivering, the doctor attending him became more concerned. She thought he was getting hypothermic and sent for his sleeping bag. A nurse arrived at our shelter to look for it, but this was the last thing he needed.

After two hours in the camp I was beginning to feel better and had just cooked some supper. Hearing of Richard's predicament, I left the meal and went along to see if I could help – more with a mind to make encouraging noises than to instigate any medical remedy. When I saw him, I was horrified. From several years' research on survival I knew the dangers of heat all too well. A shivering man in a hot environment rang all alarm bells, and instead of wrapping Richard up in a warm sleeping bag, I quickly took his body temperature. Although only an under-arm reading, the thermometer rose above 40°C, and it was nearly three hours since he had finished the race.

I called for water urgently. It would have been better to immerse him completely – not in a cold bath, which would have shut down skin blood flow further, but in cool water that would extract a lot of heat yet allow blood to continue coming to the surface. Here, however, a bath was out of the question. All we had were the bottles of mineral water dispensed for drinking. They would have to do. While Richard complained bitterly – mumbling about being cold from chattering lips – we soaked him and fanned him as best we could. It would be touch or go. He really needed hospital care, for with shivering putting heat production up to many times the normal, and sweating and other skin losses shut down, breaking the vicious cycle of heat overload can prove impossible. Sometimes it is necessary to paralyse and ventilate such a patient to stop them making more heat, and even then the outcome can be disastrous – an overheated brain may never recover from the insult. Here, as we tried our meagre best, I feared we would be unsuccessful, but then things changed. Richard began to complain of the heat and started to sweat. I sighed with heartfelt relief.

★

We all wanted to complete the last day's run as fast as possible. It meant additional pain for feet, legs and mind, but through it all the green band and the white houses drew closer, and we knew that there food and rest awaited us. We had worked hard and were about to be rewarded.

To make the finish more dramatic, the organisers set the race off in three groups at fifteen minute intervals, with the slowest first. I was in the middle group, and so caught up with others and was caught up myself. Here I saw runners from previous years handing on a fine tradition. Whether passing or being passed, it was convention to reach out and touch hands. It re-emphasised the feeling of unity that had been so much a part of the race.

About two miles out from the town, we reached a green field and then a river. The river was broad and fast, carrying water down from the Atlas snows, the source of irrigation for the swathe that cut through the desert colours. There was a bridge, but none of us seemed to find it. Instead the line of runners plunged into the water, wading across waist deep, encouraged by the whoops of local children who had come out to wave us in. The water was cold as ice but oh so lovely. We had run without washing for a week and were filthy and sore. It was a great sensation.

Beyond the river we entered narrow unmade lanes that wound through more fields and passed scattered white buildings. Then into the town proper, and we began to climb one last lung-bursting, leg-crunching hill. At the top, from a roundabout filled with flowers, the main street ran a mile or so towards a blue-tiled, domed mosque. Both sides of the street were lined with people – men, women and children – who clapped and cheered us on. I doubt that I or many of the others have ever run the final mile of a long distance race so fast. Psychology is what it is all about, and it would be difficult ever to feel better than on that final few hundred yards. As we reached the finishing arch we were met by a band playing strange pipes and drums. We cried with joy.

*

The race was over and there was a great feeling of satisfaction, but it was not long before the joy faded and was replaced by depression. The death of Mauro cast a long shadow over us all, and I think we all felt that it was time for the Marathon of the Sands to be laid to rest. To lose someone while climbing or motor racing seems acceptable. The participants in these sports make their choices while aware of the dangers. Running races should be safe, and nothing was worth the loss. As we left Morocco and headed for our respective nations, the thought of Mauro's body buried in the sand destroyed most of what we had gained from our participation. It was with sadness rather than satisfaction that we finally reached home.

Then, after being back in England for ten days, I heard some extraordinary news. Mauro was alive, unwell and in hospital, but expected to survive. The news was from the press and I simply could not understand it. His survival after so long did not seem possible. It transpired that after wandering for nine days he was found in Algeria, more than sixty miles from where he had disappeared. For me, the story remains incomprehensible. How did he get to Algeria, south of where he was lost, without noticing either the main tarmac highway or the border fence and its associated tracks? Stranger still, how do you move for nine days in the Sahara without water?

Mauro claimed to have lived on plants and drunk his own urine, but the former would provide little hydration and the latter could only make things worse. While in no way wishing to cast doubt on his story, I can state with confidence that when the body gets dehydrated, it becomes phenomenally good at conserving water. Within hours, the urine is so concentrated that to drink it can only do harm. The salts and metabolic products within it would immediately demand fluid to flush them out, and every litre drunk would need more than a litre of water to get rid of it again. It cannot work. If you do ever have the urge to drink your own urine – don't. For that matter, just as Richard resisted the idea in training, don't drink anybody else's either!

SEVEN

★

The Heart of Darkness

Antarctica . . . Not those summer months of my crossing with Ran Fiennes but mid-winter on the earlier 'Footsteps of Scott' expedition. Roger Mear, Gareth Wood and I were ten days out and struggling through frozen darkness, relieved only by the writhing serpents of luminescent green aurora. The sledges we towed behind us weighed just 50 kilos (110 pounds), nothing like as much as I would pull on the later trans-Antarctic trip, but they felt just as heavy. It was so cold that the snow underfoot was more like non-slip sand and every step we took drained all of our effort. In eight hours of pulling we could not make eight miles, and we were utterly exhausted.

Only seventy-five miles separated our winter hut at Cape Evans from our goal at Cape Crozier, but that was still far away. So slowly were we moving, we became anxious about running out of food for the return journey, yet none of us felt like turning back. The Cape was not just a point on the map; it was the home of the Emperor penguins – a place steeped in Polar history.

Although the pulling was hard, it was at the end of each day's efforts that our real difficulties began. The work of hauling kept us warm, and with warmth we were safe. Slow down or stop and the most bitter weather on earth was out to get us. The Antarctic night can freeze your skin and chill your body in just minutes if you have no shelter. Ours was the tent, but it took time to erect it and tie it down securely enough to withstand a big storm of gale force winds, sweeping across the ice-shelf and carrying all before them.

By the time we had finished pitching camp and dug snow to pile on the tent valance, all warmth we had generated while pulling the sledges had drained from our bodies and was lost to the Polar night. Speed was the key to survival. Inside the tent it was not warm but at least it was tolerable. If the stove was on and we kept the pressure lamp going, we could cook, eat and rest, and while we did so our clothing would begin to defrost. We had to brush it down before the ice melted. Our garments would be rigid in the morning, and it would be a struggle to get them on.

Morning is hardly the right word. Morning denotes sunshine or at least some light to lift one to a new day's work. Here, two months after the sun last set, only our watch alarms told us that it was time to rise. It was pitch dark when we stirred and showers of hoar frost fell from what had become overnight a crystal dome. The vapour from our breath had frozen, precipitating on the tent fabric in myriads of growing geometric forms. As they fell, they went down the back of our garments and added to the chill of awakening. It was so cold that our hands would get frostbitten while just lighting the lamp. Crawling from our sleeping bags and into more clothing was another race. But then the tent would warm up a little and breakfast could be taken at leisure. It often was. None of us were keen to take the next step. The time between exiting the tent and getting away with the sledges would be ten minutes. It was ten minutes too long.

Exposed to temperatures of below minus 60°C, even our thick down duvet jackets were useless. Only after we started the heavy work, would the warmth generated by our muscles offset the heat sucked from us. The tent had to be cleared, swept out and then collapsed, and our sledges packed up and made weather-tight. Perhaps worst of all, we needed to answer calls of nature – an awful experience in the bitter darkness. Gloves were a hindrance, but if we removed even one of our mitts to seek dexterity for a knot or a zip, our fingers would begin to freeze in seconds. I could feel the ice crystals form within them, and even if I was careful, by the time we were ready to depart, both my hands and feet would be screaming with pain. Sometimes beyond pain, an anxious thirty minutes of numbness would follow before we knew if they were actually damaged. Each of us sported several blackened and blistered fingers and toes.

When we set off, it was not entirely dark – our 'day' was timed to have the moon as our companion. It shone with glory, and its pale grey light added to the auroral flickerings above, and both were reflected from every ice surface. They allowed us to see something of the way ahead, although there was not much to see. Out in the 'Windless Bight' – the huge bay that we were crossing – the shores of Ross Island were just a dark shape rising ten miles to the north. We travelled on a flat and featureless ice-shelf that stretched hundreds of miles from here to the Trans-Antarctic mountains which were all that stood between us and the South Pole. From that direction, the vicious storms might come – the winds that had so nearly killed our only forbears.

*

Humans can withstand the coldest of environments as long as they continue to move and are well clothed, but to go out in the Polar mid-winter is to expose yourself to the severest conditions our planet has to offer. The fact that we were there at all provides an example of mankind's enormous flexibility. How did we cope with such conditions?

Africa 100,000 years ago was a place that was mainly warm and was certainly never cold for prolonged periods. The ancestral populations that moved from there to areas such as Europe must therefore have been poorly adapted to the cold, and survived by using behavioural means such as clothing and lighting fires. Today we use essentially the same techniques. Even though humans reached the far north and cold climes in pre-history, it was only in the last ten thousand years or so. Just as there has been no time for any race to lose their evolutionary adaptations to heat, there has been even less time to develop new physiological defences against the cold. That is not to say we have none, but if we go to extremes we need our intellect to extend them.

*

Apsley Cherry-Garrard's classic book, *The Worst Journey in the World,* is one of the greatest Polar epics ever written. A week after mid-winter in 1911, three men from Scott's expedition set off to seek the eggs of the Emperor penguin – a bird that was then a mystery to science. The Emperors had been discovered by members of Scott's first expedition in the spring of 1903, when they had travelled along the southern side of Ross Island in twenty-four-hour sunshine. At Cape Crozier they found penguins of a size that they could scarcely believe, but although they collected a few adult specimens to take back to their base, they could find no eggs or newborn chicks. All the young were already many months old, and Scott's men were baffled by this timing. The other more familiar and smaller species of penguins were only just hatching their eggs, yet the chicks of these enormous birds were already fully fledged. It seemed impossible, but they must have hatched out in winter. The colony must stay in the Antarctic through the months of darkness when, with the sea thickly frozen for hundreds of miles, there would be no food available. How could they survive?

The three men of the 1911 winter expedition aimed to find out, but the journey became a living nightmare. The cold of the Antarctic winter is scarcely imaginable, and as they walked through the darkness, they were wearing little more than English tweeds with an extra woolly. Their suffering began quickly. Their clothing rapidly accumulated water from their breath and this then froze. Even their reindeer sleeping bags turned into rigid frozen boards, and their nights became as bad as their dark-filled days. But it was not just the inadequacy of their garments that was to bring them close to death. The Antarctic is the windiest place on Earth, and it was the teeth of the wind that so nearly closed around their hearts.

The outward journey went fairly well, although they too were surprised by how hard it was to pull the sledges, and it took longer than they had anticipated. Over a couple of weeks they made only slow progress across the 'Windless Bight', taking the shortest route eastwards rather than following the line of the coast. Then they picked up land at a low rocky point and knew that it would not be far to Cape Crozier. To escape from the confines of their tiny tent and create a shelter that they felt would be stronger if the winds came, they built an igloo – not made of ice but formed from a low circular wall of rocks with snow pushed into cracks as mortar. A tarpaulin was used for the roof, spread across the stones and anchored down with heavy rocks. Inside it seemed both strong and weather-proof, and they made their plans for the trip to gain the eggs. Outside they pitched their tent to give them extra room.

The next day they left the igloo and their sledges and travelled straight across the base of the Cape to reach the place where the penguins had been found back in 1903. In complete darkness they found the Emperors once more, but not what they had expected. Instead of a mix of sexes, only the males stood in groups on the ice, backs to the weather and huddling together for warmth. The outer birds were obviously at great risk, but as the explorers watched, they saw the strength of altruism. After a few minutes exposed to the worst of the elements, the outer animals moved back into the middle of the huddles to recover, while other birds accepted their turn in the breach. And all the time that this went on, each bird carried an egg upon its feet, holding the precious future of their species up and away from contact with the ice, insulating it with a warm flap of fat and feathers.

The men made their unique observations and returned to their shelter, taking with them a few of the precious eggs. During their journey back, the weather began to break and visibility became so poor that they were lucky to find their igloo before the full force of the storm hit them. The hurricane-force wind attacked with violent fury from the mountains across the ice-shelf. As the hours went by, the roof tarpaulin of the igloo began to pull away. Outside, their tent was also ripped from its guy lines and disappeared into the maelstrom. The men could do nothing. They huddled deeper into their frozen bags, trapped by cold that they knew could not be faced. If they got out and tried to repair the shelter or find their tent, it would kill them in no time. They were still lying helplessly when, with a final wrench, their roof departed. They now had no shelter at all for the long journey home and, as they lay shivering in the continuing furore, each man must have thought that he would now perish.

Yet instead of death, the outcome was a miracle. The storm abated the following day and the men decided that, rather than die where they lay, they would go out and search for their shelter. Far from their search becoming a passage to oblivion, it was only a short time before they literally stumbled across the tent in the darkness. Instead of being blown far out to sea, it had been caught on an icy outcrop less than a mile from where it had been torn away. The worst the wind could offer had not freed it again, and once more they had a shelter.

It was not to be the end of their troubles, for they had little food and fuel remaining, but they survived and brought the eggs back to their hut. They had probably gone through more cold than anyone who had ever lived to tell the tale, although two of the three were then to die when they accompanied Scott on his fateful journey to the South Pole during the following summer.

Environments as hostile as Antarctica cannot be faced without danger, yet the winter journey proved that man can travel under the worst conditions our planet has to offer.

*

As with the heat, temperature regulation in the cold is under the control of the hypothalamus, the thermostat of the body, acting when cooling occurs to diminish heat losses and increase heat production. Many of the cold sensors lie in the skin, adjacent to the heat sensors, and they too are much more sensitive to change in temperature than to the temperature itself. A rapid change sends the most intense signals to the brain, explaining why the toe-dip in the swimming pool gives the strong message: 'This is really cold'. It is exactly the same as the toe in the bath warning, but this time leaving you wandering around the poolside, trying to look casual rather than gingerly advancing into the tub.

Perhaps embarrassed, you then pluck up the courage to jump into the pool, where once again you are given a dramatic demonstration of the signal strength resulting from rapid temperature change. As you splash into the water, heat is conducted so fast that skin sensors all fire wildly. The hypothalamus interprets this as water so near to fatally cold that a second message is passed to consciousness along the lines of 'This is bloody FREEZING!' You have to fight the desire to get straight back out and run for the changing rooms, but bear with it for a minute or two and all changes. The temperatures of the skin and pool tend to equilibrate and, even though the absolute temperature of the sensors is now lower than ever, their frenzied firing dies away. The water feels comfortably warm and the fuss is over. With almost no hint of malice, you call to your friends to 'Come on in, the water's lovely.'

The hypothalamus is not solely reliant upon skin signals. It also receives impulses from deeper sensors elsewhere and some of its own cells register the temperature of the blood flowing around them. It is the integration of these inputs that informs consciousness whether we are cold and triggers both behavioural and physiological responses.

One of the first things we do as we begin to grow cold is to start moving around. We either get on with the job at hand or add pointless vigorous movements to our actions, such as jumping up and down. Alternatively, we alter our clothing insulation, although the interactions between environment, clothing and activity are remarkably complex.

The effectiveness of your clothing is usually at its best when you are stationary, for in addition to the protection provided by layers of air trapped within the garments, insulation also comes from the layer of still air over the outer surface. Nevertheless, keeping still is rarely the answer in cold conditions and even at a modest plus 5°C with no wind, the insulation needed to stay warm when completely still is equivalent to about five times the average set of indoor working clothes – impractical to wear in most circumstances.

When you start moving, cold air is forced both over and through your clothing and much of the insulation is destroyed. However, this problem is usually more than offset by the increases in your heat production from working muscles, especially as cold weather clothing restricts movement, and walking over snow is also especially taxing. Well wrapped up, you often find yourself too hot, and this has even been demonstrated by men living and working in research bases in Antarctica. Overall, in contrast to the enormous amount of clothing needed to keep you warm while stationary at plus 5°C, the equivalent insulation of normal office dress will do the job at minus 18°C if you are working moderately hard.

Of course, there is more to the weather than the temperature alone and more to clothing than just insulation. During a stay in Antarctica in the 1950s, Admiral Byrd invented the concept of windchill, with which we are now all too familiar. His experiments were simplistic in the extreme, most of them involving observations of the time it took for empty baked-bean tins filled with water to freeze under different wind and temperature conditions. Still, the results were striking. For example, a temperature of plus 5°C with a stiff wind of 25 m.p.h. is equal in cooling power to a still air temperature of below minus 50°C.

The use of such windchill figures can make cold windy days sound highly dramatic, and they are widely quoted in modern television weather broadcasts. In reality this is ridiculous. We behave differently to baked-bean cans, not least because we usually wear clothes on cold windy days. These protect us from some of the additional cooling, and so to take windchill as the effective temperature is as foolish as ignoring the effects of wind altogether. During our Antarctic winter journey, the windchill temperature sometimes fell to less than minus 100°C on days when winds were combined with still air temperatures in the minus forties. Yet it did not feel half as cold on those days as it did in black, fog-filled stillness when temperatures dropped as low as minus 67°C.

At the same time, wind effects cannot be ignored. Exposed hands have fingers that behave more like bean cans than the rest of the body, and a windy day in Antarctica will freeze bare skin in a matter of seconds. The reality is that the cooling power and dangers of the environment are somewhere in between the still air temperature and the more dramatic windchill. You must bear this in mind if you are to operate safely.

*

Through the seventy-five years that followed Cherry-Garrard and his two companions' *Worst Journey,* nobody repeated any significant Antarctic winter venture. Then, in the autumn of 1985, we arrived at Cape Evans with the 'Footsteps of Scott' expedition. As with Scott's 1911/12 venture, the overall aim of the 'Footsteps' team was the journey to the South Pole in the long Polar summer of the following year, though the freezing of the seas around the continent had forced us, like Scott, to come down to Antarctica early and spend the winter waiting. Living in a hut through the winter on the same beach as Scott's team had done, it was inevitable that we should attempt to repeat their mid-winter journey. It would be both an historical re-enactment and an ultimate test of the preparations for the coming South Pole trip. If you can live, work and travel in the Antarctic winter, you can do so in any circumstances.

The three of us had therefore set off from our winter hut at Cape Evans on the same day of the year as the original trio. After travelling south and crossing Discovery point, we had gained the ice-shelf to strike eastwards out and across the Windless Bight. Now, in the moonlight ahead, we could just discern a low line in the darkness. It was the rocky promontory of history, and we knew that there we would find the remnants of the circle of stones in which our predecessors had faced their mortality. We approached in awed silence.

*

Modern thinking about clothing has gone down the road of using multiple layers to achieve protection, with flexibility to cope with a variety of activity and weather conditions. The addition of water to the cooling equation can be as dramatic as the addition of wind, since water evaporation from the surface will extract enormous amounts of heat, and any water penetrating the garments will much reduce their insulating capacity. In order to survive in bad weather, we usually need windproofing, water proofing and insulation, and a huge range of garments has been developed to meet these requirements. The choice can be bewildering when deciding what to wear, especially if you plan to walk through the Polar night.

In the Antarctic, as in other very cold regions, rain or melting snow is not usually an issue; it is actually drier there than in the central Sahara. Clothing need not therefore be waterproof in the way it would be for more normal winter conditions. Even so, while pulling sledges, our bodies would give off water in both sweat and breath. There was therefore a risk of condensation and wetting of the clothing from the inside. This had been a source of terrible problems during the original journey when their clothing became rigid. To avoid the same difficulties, we had our garments specially made to be waterproofed on the inside. Our sleeping bags were similarly protected, and even one of our layers of socks was waterproof, protecting our boot insulation from the sweating of our feet. The proofed clothing also allowed us to choose duck-down for insulation which, despite modern innovation, remains the warmest material for weight as long as it stays dry.

*

It was colder than ever and our thermometer had gone off the scale. After our return to base, we would hear that temperatures had dropped to minus 67°C, but now we only knew that it was dire. It was difficult to make things out as we drew nearer to Igloo Spur, for along with the darkness we were enveloped in thin mist. We guessed that it must have been emanating from an open tide-crack on the other side of the

Crozier peninsula, for nowhere else would there be a water surface to generate it. In such cold conditions, mist must be a rarity indeed.

Our plan was not to stay at the igloo, but to move on to Cape Crozier, leaving two out of our three sledges behind. The way ahead was largely over the rough rock of the Cape and it would be difficult to pull the sledges without damaging them for the forthcoming South Pole journey. The third sledge was a rough and hardy pulk made of indestructible moulded plastic and equipped with metal runners. The three of us would go on pulling it together with just a few days' supply of food and fuel. But before going on to the Cape, ten miles away, we wished to visit the shrine of Polar exploration. And so, as we reached Igloo Spur, we set up camp in order to warm up both ourselves and our cameras.

The way to the igloo was down a slight slope and we kept to the shallow ridge line as we descended. The low stone circle was only a couple of hundred yards away and we found it easily, spending an hour there photographing and filming. It was a moving experience but it left us chilled to the bone. By the time we set off back to our tent, we were badly in need of shelter, the stove and a hot drink.

As we walked back up the ridge it became quite difficult to see anything. A slight breeze had risen and more mist had been blown across the Cape, but the rock spur was only a few yards wide and we felt confident that we could not miss the tent. We were wrong. We found nothing, just an empty snow slope above the spot where the rocky ridge petered out. Puzzled, we turned back and made another sweep downwards, but once more to no avail. We did not even return to the igloo but went right down to the end of the rock and then on to the flat ice-shelf without coming across any familiar features. Obviously the layout of the land was more complicated than we had thought. We paused and considered our strategy.

With our head torches we could see each other when we were quite far apart. The obvious thing to do was to form a broad line and then sweep back up the hillside so that we could not miss either tent or igloo. We spaced ourselves about fifty yards apart and climbed once more – with no success. The mist was getting thicker and it became hard to discern rock from snow and up from down. It seemed impossible that we should not find our camp. Using a compass to move at right angles to our first line by about a hundred yards, we spread out again to work back down the slope. Still no tent, no igloo, and this time no rock. We were truly becoming disorientated, and we each began to feel rising panic, but self control was the only answer. If we were logical, we had to be able to find our things even if it took some time. We had to quarter the whole area, making the search square bigger and bigger until we met success.

It was now obvious that we had made a fundamental mistake when we first left the igloo. Perhaps it had something to do with the reading of the slope, which can be notoriously difficult in whiteout conditions. If we had taken a spur off the promontory as we headed back towards our camp, that might have led us in any direction. The best thing to do now was to ignore any pre-conceptions about the terrain and to simply search in an organised fashion. The tent could not be far away.

In the end, we quartered the ground for the best part of another hour of rising fear before there was a sudden cry from Gareth.

'The igloo – I'm at the igloo.' Roger and I hurried towards his bobbing light.

He was right, we were back at the stone circle and the tent was close if we took the right ridge line. We then found our camp quite easily. Our mistake must have been made through lack of concentration. It is often such mistakes that kill.

*

The choice of clothing for our winter Polar journey would not be effective in less extreme conditions where a good outer waterproof layer is needed to stop rain or melting snow from getting in. This can be either a totally impermeable fabric, keeping out both water and water vapour, or a vapour permeable material such as *Goretex*.

The vapour permeable waterproofs have microscopic pores, large enough to let water vapour pass, but too small to allow liquid through. Their aim is to limit condensation inside the garment building up from vapour coming off the body. When working hard, inner garments may become as damp from condensation as they would if it were raining and you had no waterproof. However, vapour permeable garments are expensive and even the best will not let much vapour through when it is cold. At low temperatures, water vapour condenses or even freezes when it first meets the inner surface of a jacket and, once condensed, it will no longer be able to penetrate. In winter cold, the specialist vapour permeable jackets are really little better than their cheaper non-breathable alternatives.

Less extreme cold also influences the best choice of insulation. Down is useless when wet, and so is best avoided unless in conditions where all water is guaranteed to be frozen. If not, duvet clothing and sleeping bags are better filled with modern synthetic materials which, although heavier than down, are fairly warm when wet and dry out quickly. Artificial fleece jackets also absorb less water than woollen jumpers, and polypropylene underwear does not become damp like its older cotton counterparts.

Another important consideration is what to wear on your head. When you are well wrapped up in cold conditions, as much as ninety per cent of your remaining heat losses can come from your scalp, and so putting on or taking off a hat can change thermal equilibrium substantially. Without the problems of fiddling with buttons or zips, you can regulate you overall comfort and may even influence thermal balance in distant parts of the body. If you have cold hands, you should put on your hat.

*

After finding our tent, we thawed out and ate a meal, but soon it was time to move on. Despite the fact that we had already been up and out for more than twelve hours, the Igloo Spur was far too exposed for safe camping – as our predecessors had found to their cost. We packed up, anchored down two sledges that we would pick up on our return, and set off pulling the rough pulk between the three of us.

On the broken ground we kept stumbling into each other and, cold and fatigued, none of us were in a mood for tolerance. Tempers frayed as we argued about the best route and we spent some time quite lost. The first sign that we had reached the seaward side of Ross Island was ground littered with the corpses of frozen penguin chicks. These were not Emperors but the smaller Adelies which dwelt in the summer months on the hills immediately behind the Cape. It was difficult to believe that the Emperors could really be here in the heart of winter's darkness until we heard through the darkness eerie, distant, echoing calls.

We stayed the night on the Cape before looking for the penguins in earnest. Armed with head torches, we descended from the land and searched along the cliff where the ice-shelf met the frozen sea. It was a difficult task, for the shelf was riven with deep cracked inlets, in any of which the birds might be sheltering. After hours of searching we found, not a host of birds, but just a couple of small groups standing in the darkness. Although few in number, I was not disappointed when I saw them. Disturbed by our approach, they turned and dark backs which had faced the deep cold were replaced by white breasts that shone in the light from our torches. From their curved stiletto beaks rose haunting melancholy calls which echoed from the walls of the ice-cliffs behind them. We were in the court of the Emperors of Antarctica and the trumpets had sounded. It was an extraordinary moment and a great privilege to be there.

*

Our physiological responses to cold are limited to two mechanisms: the first involves cutting down heat losses and the second increasing our heat production. Heat losses are reduced by shutting down warm blood flow to the skin which limits convection, conduction, and radiation. Furthermore, a simultaneous decline in the blood flow to deeper layers of fat, and even to superficial muscle, helps to prevent heat being lost from deeper, warmer regions.

The insulation we gain from the fat under our skin is one of the most unusual aspects of our evolution. All our primate cousins – near and far – rely on warmth from fur, and there is considerable debate as to when and why we went down a different road. Conventional wisdom suggests that the change improved our heat tolerance, since bare skin permits efficient heat loss. On the other hand, fur grants considerable protection from the rays of the sun in hot climates and if it had no advantages, other tropical animals would also be bare skinned.

A striking alternative suggestion is that we developed sub-cutaneous fat as an adaptation to life in the water – an idea fundamental to the 'marine theory' of human evolution. I have mentioned previously the possible drives that brought our ancestors to their feet – the advantages of reaching for fruit, protection from the midday heat, or the need to move fast and efficiently over open plains. At first sight, all these suggestions seem credible but the development of upright posture must have been a slow process, and any change in anatomy or behaviour can only evolve if it confers significant survival advantages at every step on the way. This calls into question the conventional explanations of our uprightness. If becoming fully erect took many generations, there must have been a prolonged period when our ancestors were only partly erect – a posture which would grant few of the advantages that being upright would eventually confer and many disadvantages. Moving half bent while losing your ability to move on all fours would seem like a recipe for disaster.

The 'marine theory' of evolution attempts to get round these problems with a radical re-interpretation. It suggests that instead of a move on to the savannah dividing the common ancestors of chimps and man, the separation was much more definitive – a change in sea level around five to seven million years ago that led to our own ancestors becoming trapped on an island located roughly where the southern uplands of Ethiopia are today. As a result, our ancestral group spent much of the time living in and around the water and, when danger threatened, they simply walked out as far as they could go. They would also have spent long periods wading in search of seafoods, and so had good cause to come to their feet. The theory is attractive because it can cite progressive advantage to the change of posture, especially as the buoyancy of the water would have helped us not to fall over when we were still less than good at walking upright.

An integral part of the marine theory also suggests that, if we spent long periods in the water, we would need protection from cooling, and that the development of sub-cutaneous fat insulation was effective for this. Certainly the only other mammals that have a skin structure similar to our own – whales, manatees, hippos, and elephants – are either fully aquatic or spend a great deal of time immersed. Yet, however tempting, the ideas remain highly contentious, not least because there is no real fossil evidence to support it.

Whatever the origins of the fat layer beneath our skin, the insulation it provides is very dependant upon its thickness. Due to hormonal influences, females tend to have more than males and this may reflect a different evolutionary need, although not necessarily for its insulating qualities. It may be that women may have bigger sub-cutaneous fat stores to help see them through the demands of pregnancy. Nevertheless, it raises an interesting question: would women be better suited for Polar exploration than men?

This proposal was once made by a female correspondent writing in response to a newspaper article I had published on the science of Polar travels. She pointed out that, despite all my interesting observations concerning hardships, women might have done better. Not only were their body fat stores larger, but they had a greater resistance to discomfort and cold. She qualified the point, however, with one penetrating observation: '. . . since all Polar journeys are prime examples of the heroically daft,' she wrote, 'what makes me think that where men will see the heroism, women will only see the daftness?'

Increased heat production in the cold occurs chiefly through higher levels of activity but, if stationary, we can turn to shivering. When we shiver, the brain switches muscle fibres on and off at very high rates so that there are a lot of brief uncoordinated contractions. These produce no useful work but do burn fuel and produce heat. The trigger is usually a cold skin, especially in rapidly changing temperature situations. This is quite different to heat regulatory changes which respond mainly to a change in core rather than skin temperature. Whereas sweating does not start until the body has overheated to some extent, shivering can take place immediately, even in quite modest conditions.

Sitting out on a summer's evening and feeling quite warm, you may even shiver if the sun is suddenly hidden by cloud, although there has clearly been no fall in your core temperature or any significant heat loss from your body. The response is made solely to the rapid change in skin sensor temperature, although your core warmth is not entirely irrelevant. It exerts a strong inhibitory action on the shivering process, so that if your core temperature is normal, you can never shiver vigorously. The summer's evening experience is mere passing *brhhhhh* . . .

The heat generating capacity of shivering is fairly limited. Even with teeth chattering and body shaking like a leaf, you only increase your resting heat production by about five-fold, taking you up to around 500 watts – rarely enough to keep your whole body warm. Perhaps the evolutionary purpose of shivering was, as some experts suggest, less for the production of heat than to prevent resting muscles from becoming too stiff with cold. This would allow our ancestors to sleep through the African darkness, with core temperature dropping but still with muscles plastic enough to face a nocturnal threat. Although this is an attractive hypothesis, I am not sure that it stands up to closer scrutiny. As we begin to cool, shivering always starts in the muscles of our trunk – the ones that give us the *brhhh* in the upper chest and shoulders. Only later, when core temperature really drops, do the muscles of our arms and legs become involved, and by that time the shivering is so violent that one could never rest anyway, let alone sleep. To me, this suggests that the main teleological purpose must be to limit falls in core temperature rather than to maintain limb suppleness. Its ineffectiveness can be attributed to the fact we evolved in climes where it did not need to be efficient.

*

Early in 1979, Major Mike Kealy D.S.O., based at the Special Air Service headquarters in Hereford, was appointed to head up a squadron of elite armed forces performing counter-terrorist operations in Northern Ireland. He was an experienced and well respected officer, used to being out on winter hills and mountains, but his duties had just taken him through a period of easy desk-based work. Having lost some of his previous fitness, he needed get back into good physical shape and decided to go on part of the course that the SAS uses to select new recruits.

Only the toughest and most experienced members of British Army units can apply to join the SAS, and the selection course is run over a period of five weeks in order to whittle down 150 applicants to perhaps a dozen or two. These then face a final gruelling week, and although all the men who remain are the hardiest of the hard, many fail on the final hurdle. The last exercise of the course is known simply as 'Endurance', and it entails carrying backpacks of more than 25 kilos across nearly fifty miles of rugged and difficult mountainous terrain at a time of year notorious for bad weather. The potential recruits must move as fast as they can, and sometimes nobody is deemed to have completed the test satisfactorily. If that happens, nobody joins the SAS that year.

Satisfied with his performance on some of the earlier test exercises, Mike Kealy thought that he too would take on 'Endurance'. If he did well against recruits that were almost half his age, it would confirm that he had come far and was truly back to strength. So, at about one in the morning, fully dressed for the hill, he joined the others for breakfast – a massive, traditional English fry-up of eggs, sausages, tomatoes and bacon. The meal completed, they all got into the back of a truck and were driven to the foot of the Brecon Beacons.

The Brecons are not big mountains but they are bleak and very exposed. That morning, high winds were hurling snow across cloud-covered tops. The men set off individually, with ten minutes between them, beginning soon after 2.30 a.m. First they climbed through the darkness towards higher ground and, as they gained in altitude, the wind rose and the temperature fell. Soon it was blowing at more than 60 knots, and the thermometer was down to minus 9°C. The combination produced a wind-chill well in excess of minus 50°C.

Everyone found the first couple of hours incredibly hard. The route took them off all paths and up through dense tussocky grass, which takes a heavy toll on strength even when bare of snow. The work, however, did have compensations – nobody felt particularly cold. At 1,500 feet above their start point, the going became easier as the ground levelled out. It was here that the conditions began to take effect. For men working less hard, the environment became a vicious and dangerous enemy, and everybody began to suffer terribly.

Nobody knows exactly what happened to Mike Kealy. At around 5.30 a.m. he met and spoke to a number of other recruits who were briefly taking shelter beside some rocks. They were contemplating dropping down from the high ground to take longer more sheltered routes, but Mike exhorted them to go on. He obviously felt that crossing the high Brecon was still possible, and certainly he should know for he had done it many times before, and even in similar weather conditions. Every foot of the way was familiar to him, and furthermore he had been an instructor on the dangers of bad weather and hypothermia. If anyone would be safe, it would be him.

As the night turned into an equally savage day, Mike Kealy's body was found just a few miles further from the point at which he was last seen. It was 9 a.m., and he was resting against a rock with some of his clothing and equipment discarded nearby. It looked for all the world as if he had simply sat down and waited for his fate. The formal enquiry discovered nothing more. The only lesson to be learned was one that had been heard many times before – even the fittest and most expert can be overcome by cold.

*

If you choose to face hardship, your clothing, behaviour and physiology cannot always be relied upon to protect you. Throughout history men and women have become victims of excessive cooling, often due to a dangerous combination of cold, wind and water rather than especially low temperatures.

If the body fails to maintain core temperature, a number of things happen both consciously and subconsciously, depending upon how cold you are. When core temperature first begins to drop, you consciously feel chilled, and this is a useful signal to prevent further deterioration. It makes you actively try to improve your protection, to move around more to generate warmth, or to get out of the situation, and while you do so, your body cuts down on surface blood flow and sets you shivering. These reactions occur at core temperatures of between the normal 37°C and a low 34° or 35°C, and are usually enough to stop things from getting worse. Often you will over compensate and end up feeling hot.

Sometimes, however, these defences will not be enough and cooling will continue. Then some rather strange responses take place. First, your thought processes begin to fail, with some individuals becoming withdrawn and uncommunicative, while others exhibit wild disinhibition. The variation seems to be analogous to getting drunk, when one might end up either drowsy or overactive. The difference is down to whether the essentially sedative action of either cooling or alcohol knocks off the centres that inhibit innate boisterousness before they knock off the centres that keep you alert and active. Unfortunately, neither being dopey or irresponsible is helpful in a cold survival situation. Both serve only to increase the level of danger.

If things get worse, further changes make it even harder to recover. At core temperatures of around 32° to 33°C, hypothermia victims begin to stumble or stagger and the mild personality changes exhibited earlier become aggression, confusion, or pre-coma. Any thought of saving themselves disappears at this point, and sufferers often exhibit particularly strange behaviour where they try to undress. If asked what they are doing, they will often say that they feel too hot, and this reflects a change of great seriousness. The hypothalamus is packing up and misinterpreting the situation in a manner analogous but opposite to that seen when the body is overheated. It begins to tell consciousness that heat is the problem and, simultaneously, it turns off both shivering and the insulating blood flow responses of the skin. The victim is left near defenceless and rapidly heading for the point of no return.

Once a casualty has lost hypothalamic control, he or she has entered a vicious downward cycle. At around 30°C, drift into coma begins, and while there is still some way to go to the temperature at which death will take place, unconscious, inert and with protective mechanisms switched off, it will not take long to get there. At around a blood temperature of 24°C the heart stops.

Every year sees fatalities of this nature, even in Britain where it is never exceedingly cold. People ask me for advice about what action they should take if they are with someone who is becoming hypothermic. It is a difficult question to answer. The correct, perhaps lifesaving action in one situation may be entirely and fatally wrong in another. When someone in a group starts getting cold, you must always act quickly to increase their clothing insulation. After that you are faced with a more fundamental decision. Do you stop or do you go on? The problem is the balance between the victim's heat production and heat losses. You need to try to keep up the first while minimising the second, but the two aims are incompatible.

Most hypothermia victims out walking or climbing in Britain will not be so far from civilisation that you cannot guide, encourage or push them to proper shelter and warmth. To my mind, you are therefore asking for trouble if, at the first sign of hypothermia in a member of a party, you decide to lie up in temporary shelter and perhaps spend a night or more waiting for rescue. In most cases, it would be better to keep on moving. This advice runs contrary to that often given for, generally, it is recommended that if one person is too cold, the rest should halt, keep the victim as warm as possible, and send for help. I believe that groups sometimes follow this recommendation rather too literally. Inexperienced parties without full equipment have stopped when one of them has been a bit cold and then tried to shelter in awful conditions. The hypothermia victim, and perhaps some of the others, have then died when, had they pushed on, they might have got everybody off the hill quite safely.

Of course there are numerous situations in which stopping and lying up is the correct choice. If the victim is more than just cold and is actually showing signs of staggering or confusion, the situation has become so dangerous that it is probably better on balance to stop. Continuing when you don't know where you are going also tends to be unwise, and hypothermia problems often arise because bad weather has led to a party getting lost. Civilisation can be infinitely far when walking round in circles.

If forced to lie up, you must attempt to reduce everybody's heat losses to less than the meagre warmth they will be producing. This is impossible if you are ill-equipped. Ideally, one should have enough protection to cope with getting stuck out in the worst possible conditions that could be encountered but, for practical reasons, this is rarely the case. Nevertheless, all groups should carry something to make a windproof and waterproof bivouac if they are out in winter and must never rely on those lightweight, silvered survival bags or blankets. Although temptingly light and small for the backpack, they have been shown to be almost useless. They are certainly less effective than a decent plastic bag.

Hot drinks are also part of standard teaching, and these can be of great value. Once again, however, do not treat them as too important, for it is asking for trouble to sit around preparing them instead of keeping moving. Usually they are not worth the effort unless ready made in a vacuum flask. Food, on the other hand, is always of value. Even if not heated, it will rev up your metabolism and, more important, will help to sustain the supply of fuel to muscles so that you can keep on working. It will also make you less likely to develop a low blood sugar, a factor which I believe may have been important in some of the apparently inexplicable exposure accidents.

The death of an expert like Mike Kealy can be difficult to explain and makes one wonder if he experienced some unusual physiological problem. To my mind, his breakfast before departure may be relevant. Traditional English breakfast fare is sorely lacking in carbohydrate, and after many days of arduous activity, his muscles were probably low in carbohydrate when he came into the dining room and little better when he stood at the foot of the hill. By the time he had climbed the steep slope, more would have been burned through his strenuous efforts, and with depleted glycogen stores and continuing work at a level demanding carbohydrate, it seems quite possible that blood glucose levels fell. With this, he would have slowed down and so produced less heat. At the same time the low sugar levels may have affected his hypothalamus and turned off his physiological protection. Unfortunately, we can only speculate, and if we do so, we can think of other possible ways in which defences may be overcome.

Large fluctuations in heat production occur when working in the cold, particularly when walking in the mountains. Going uphill requires more effort than going down or walking on the flat, and as you climb you produce so much heat that you will often need to activate your heat loss mechanisms to get rid of it. Approaching the top may therefore see you flushed, sweating hard, and with much of your clothing in your pack. Then, as you reach a crest or summit, the slope changes, work and heat production are suddenly cut, and you can enter a period when heat losses are enormous. With your skin blood flow set at rates suitable for the climb, clothing damp with sweat, and no inclination to stop and dress up again, core temperature can plummet. Some experimental work with which I was involved suggests that the combination of a falling core with still high skin temperatures can cause some confusion in our temperature regulation.

These experiments simulated just this sort of situation in a climatic chamber. Using an artificial wet, windy and cold environment and military volunteers as our guinea pigs, we discovered that when our subjects moved from hard exercise to easy exercise, some found it very difficult to keep going, and several very fit men had to give up because they felt so unwell, despite having skin that was still warm and core temperatures, although falling fast, above normal. On the other hand, if they managed to continue, they felt better again once their core temperature had dropped to subnormal values and their skin was cold and shut down. It seems as if the abnormal combination of warm skin and hot but falling body temperature was confusing. It is easy to imagine that, if they been out on the hill, some of them might have slowed down so much while feeling awful that got themselves into trouble. This provides an alternative explanation for Mike Kealy's downfall as he finished his brisk, warming climb and switched to a tramp across the freezing tops.

*

The accident occurred after we had ditched the sledges and were struggling along with crushing backpacks. My bindings had broken some weeks before, and I was following on foot where Ran had just passed on skis. The Arctic pack ice was most untrustworthy and even here, close to the North Pole, the thickness varied from a few inches to tens of feet. The problem was judging the difference as you walked. Without my weight distributed along the length of a ski, I was much more likely to break through, and so it was with no surprise that I suddenly found the snow crust giving way as I followed on Ran's track. The shock came as I continued down. With horror I realised I had broken through into an open water crack.

There had been no sign on the surface of this fissure despite it being about five feet across. Wind blown snow had built a delicate carapace to disguise it, but now that had gone and just thin ice lay upon the dark water surface below. I went straight through and under, swallowing cold brine as involuntarily I gasped. Immediately, I popped up, buoyant with the air that was trapped within my clothing. I spluttered and wiped the water from my face. For a moment, I wondered why I didn't feel cold, but that soon changed. Within seconds, icy water poured through my clothing and a tight band grasped my chest. It was so cold that it hurt, and I knew that it could squeeze me to death. I had to get out quickly.

I shouted desperately to Ran while looking around and searching the sides of the crack for an escape route. The sides were vertical and smooth, and although I was only four feet or so below the top, it might as well have been four hundred. There was no way to climb out, and I shouted again more urgently. There was no reply. Ran had been close when I fell – stopped only twenty yards ahead – but he had been facing the other way. With his back to the wind, he might have set off walking without looking back, and I began to feel the fear and panic welling up inside me. Then clearer thoughts of self-preservation punched through the pointless whirl of fear.

I couldn't climb out with a backpack on. It was heavy enough already, and rapidly becoming sodden. I had to get it off before it dragged me down, but every time I tried, I sank as I attempted to free my arms from the straps. The cold water over my face was too much to bear as each repeated attempt became more difficult. My fingers were turning numb and they would soon lose their grip altogether. Time was running out and terror again took hold. It was then that I heard Ran calling.

'Mike, Mike! Are you okay?'

Thank God! He hadn't disappeared after all, and the urgency in his every syllable pushed back the fear once more. Then he was there, above me.

'Here, grab the stick.' He lay down and lowered a ski-pole full length. I seized it but as he pulled I realised the futility.

'I can't! I've got to get this sack off.' My voice was querulous with fright but the hope he offered gave me strength.

There was only one way to get rid of my load. I had to let go of the ski-pole and go under to fight with the straps that contained me. I took a deep breath and went for my only chance. Fumbling with the straps, I began to sink, and the cold water on my face was almost impossible to bear. I felt faint; I felt like letting go. My mind began to darken. Then, just as my lungs were bursting, I broke free. The sack was off and I pushed back to the surface, still thinking clearly enough to hold on to my load as I did so. The sack contained vital food, equipment and the radio.

'Quick! Take the sack or it'll sink.'

Ran hooked his ski stick into the hoisting strap and up it went. Then it was my turn. I grasped the stick and hauled once more, but I came only slightly out of the water. The weight of my saturated clothing was now too much, and there was no way that Ran could heave me out. I fell back exhausted within moments. My whole body was rapidly chilling and my muscles were losing their function.

'Come on, Mike. Come on!' Ran shouted.

It was all very well for him to shout. How could I come on? I could feel my strength ebbing away with every moment that passed, and within seconds there would be none left at all. Desperately I looked round again. This time I saw a chance. To my right, perhaps twenty yards away, the crack narrowed. Perhaps there I could get some purchase. If I bridged with my legs at the same time as Ran pulled, perhaps there was some hope.

'Quick, Ran. Where it's narrower.' I pointed.

We both moved along, me half swimming and half pushing off the walls. Then Ran lowered the ski-pole again. I held it with both hands and heaved once more, this time thrusting one foot on to the ice wall in front of me and one on the wall behind. I moved up and jammed my feet hard to rest for a moment. Then I repeated the action – up a few inches, move my feet, up a few inches and rest. Slowly it paid off and I gained height. Then with a heave from Ran I flopped over the lip and lay gratefully on the ice-floe – my sopping clothes instantly transformed into rigid frozen boards. My muscles also seized up from the effort and the cold and I realised that I needed the tent and warmth quickly. I had been saved from an untimely burial at sea.

*

Full immersion in cold water, rather than just wet clothing, increases the danger of hypothermia immeasurably. The waters of the Arctic ocean are close to zero, but even much warmer water will conduct heat away so efficiently that you will become cold rapidly. Keep still long enough, and you will be close to death in the local leisure centre; fall in the sea away from land and even in the summer you will succumb. Each year more than 140,000 people throughout the world die in the water. Few of these fatalities are due to drowning alone; most take place in water that is cold. People die of either cold shock when they first fall in or more slowly from the inexorable onset of hypothermia.

Cold shock is ill understood. Pain and reflex responses triggered by sudden cold exposure lead to rapid increases in blood pressure accompanied by a radical slowing of the heart. At the same time, the acute exposure to cold triggers an involuntary gasp which may make the victim inhale water, followed by further uncontrollable breathing which soon washes out carbon dioxide, and so changes the acidity of the blood. The result is a combination of metabolic upset, confusion and impending heart failure, and many cold water deaths occur with victims just a few yards from safety. Even good swimmers can be totally immobilised.

I have experienced something of this myself when I was one of several volunteers in a series of tests on helicopter escape equipment. It has long been recognised that as soon as a helicopter ditches on water, the rotating blades tend to flip it straight over, the cabin floods, and many potential survivors drown before they are able to escape. The tests were evaluating the use of tiny compressed air cylinders, designed to give about two minutes of air and so increase the chances of people getting out. They took place in a 'dunking' facility in which, strapped into a typical small helicopter seat, with cabin wall on one side and aisle on the other, I was suddenly inverted into the water and had to undo my belt, get into the aisle and exit through one of the escape doors.

It was surprisingly difficult, even though I knew exactly what I had to do, and the water was warm and light. It was very disorientating to turn upside down like that, but with the short-term breathing device, escape was possible, for it contained plenty of air for the time needed. I and all the other volunteers were successful. We then repeated the test in water at 5°C, in semi-darkness. It was quite different. As the rig inverted and water flooded in and over us, the urge to gasp was near to uncontrollable and needed to be fought as we tried to put in the mouthpieces of the breathing cylinders and trigger the flow of air. Then, once breathing, and in considerable pain from the cold on our faces, we struggled to work out what to do. At or before this stage, many of the volunteers failed and needed to be dragged out by safety divers, while the rest of us managed to make sensible decisions and to head for the lit up exit. As we did so, however, the cold water drove our respiration wildly, and no doubt our hearts were also slowed by the diving reflex. The result was a gradual loss of concentration. The air supply lasted for less than thirty seconds when breathing so quickly. Only a few of us actually managed to get out without help. I, for one, had no idea what was happening when I first broke surface.

Beyond cold shock, and again even when individuals can swim, survival can be short unless the victim is in the warmest of the world's waters. The fact is well known, and the likely time to fatal hypothermia often dictates the duration of active searching for the survivors of marine accidents. Research to identify likely survival times was pursued in earnest during the Second World War when dog-fights over the English Channel and the North Sea led to large numbers of both British and German aircraft crashing in the water. Although the planes were lost, many of the airmen baled out and survived, and because these men were valuable assets to their respective nations, great efforts were made to save them from a watery grave. Nevertheless, no one wished to waste precious resources searching for men who had already perished. It was necessary to know exactly how long they might live and swim in such seas at different times of the year. The research in Britain was strictly limited by ethics but, in Germany, some of the Nazi scientists had no such qualms. The experiments performed became the yardstick of depravity as inmates of concentration camps were dropped into cold water tanks and observed as they died. The results from this inhuman work are still used, although not without ethical dilemmas.

Some insist that data from Dachau and elsewhere should be ignored for all time, but personally I could not disagree more. If anything useful can be salvaged from the horrors of that era, I believe that it should be. At least then the deaths may not have been in vain. There is still a need today to know how long a man or woman might survive in the ocean, for planes still crash, ships still sink, and yacht crew are still lost overboard. To provide answers, we must use all available sources and, to my mind, ignoring evidence from the concentration camps is to perpetuate the crime against humanity.

The interactions between water temperature, clothing, individual body fat and physiological responses are very complex and, even using all available data sources, final cooling rates are still very difficult to predict. Modern experiments, rightly constrained by the need to avoid risks to volunteer subjects, can only observe the early phase of cooling in water. To avoid these limitations, scientists have tried to develop computer simulations of cooling to extend the available information. Some of these simulations are among the most complex 'mathematical models' in existence, utilising some of the world's most powerful super-computers. They have produced predictive tables used by the authorities to guide them when searching for victims of accidents at sea. Yet, although they give some indication of when it is time to stop hoping, even the most sophisticated predictions can be proved wrong. Some individuals survive whatever the odds against them.

A recent case, of which the whole world heard, was that of a competitor in a single-handed round the world yacht race. Crossing the Southern Ocean, bound for Australia, his yacht was capsized by a freak wave and he was trapped beneath it. His vessel was spotted by a searching aircraft several days later, but it was to be many more before a ship could reach him. With the water temperature at only a little above freezing, survival had to be impossible. Listening to news reports as the days went by, one had to wonder why the authorities were bothering.

Underneath the hull, Tony Bullimore felt otherwise. Wearing a waterproof survival suit over considerable clothing, he was also lucky enough to be overweight. The resulting total insulation both natural and manufactured, along with the fact that he could spend much of the time wedged out of the water under the boat, meant that he was not robbed of his warmth. Certainly he needed sheer willpower if he was to live, but nobody who sails around the world alone is lacking in that type of determination. More than a week after his boat was knocked down, a Royal Australian Navy vessel pulled alongside to retrieve the body. Tony swam out to meet them. It was an extraordinary case, but at least it could be explained logically. Some other water survival stories are less amenable to analysis.

*

In the early hours of a March night in the late 1980s, a fishing vessel, the *Hellisey*, was upturned without warning off southern Iceland by a combination of snagged nets and a large swell. Of the five men aboard when it happened, two were below decks as the boat turned turtle and were presumably drowned, or died of cold shock, immediately. The other three found themselves together in freezing darkness, clinging to the upturned keel. They knew that staying with the boat provided the best chance of rescue but the option was rapidly withdrawn when, after only a few minutes, the hull sank. Stranded more than three miles from the shore, with the air temperature below freezing and the water at 5°C, the three men's situation was impossible.

All predictive models would give them no chance of living, and likely survival times were of the order of ten to twenty minutes. Oddly, the models would also predict that if they swam for the lights they could see on the distant shore, their time to death would be cut further still. It is a strange fact, observed in both computer models and experimental data, that people immersed in cold water will cool faster if they move in an attempt to try and keep warm – their increased heat production is more than offset by greater heat losses from the cold water pushing through and around their clothing with the swimming action. It is a consequence of the conductivity of water being more than twenty-five times that of air.

Yet, even if computers pronounced them dead, the men themselves did not agree. Perhaps not knowing the theoretical risk, perhaps recognising that rescue wasn't coming, they struck out for the shore, swimming as strongly as they could for their chance of life. Only one, the 23-year-old Gudlaugur Fridthorssen, survived, the others lasting for about as long the computer models would have predicted. According to Gudlauger's account, it was less than ten minutes before he found himself alone in silent darkness. He then swam on, and although he admits that the pain in his legs and arms made swimming difficult, he obviously managed to work hard enough to prevent his core and muscle temperatures falling. In the end, he swam for more than six hours until, as the sky lightened and the sun rose, he found himself approaching a beach. Half crawling, half washed up by the incoming tide, he landed not far from a farmhouse and stumbled there to tell his tale.

*

Gudlaugur Fridthorssen's survival raises several questions, only some of which can be answered. The two most important factors in his favour were his size and his physical fitness. The Icelander was as big and strong as they come, and along with his very considerable muscle he also carried a great deal of fat. There is no doubt that if you wish to survive in water, the fatter you are the better. It is just about the only positive thing that can ever be said of being overweight. Unlike clothing, the insulation under your skin never becomes sodden, and it seems that he had enough to limit the rate of heat loss even when he started to move and the cold water swirled around him. His warm woollen fishing kit would have also helped somewhat, as would his huge muscle bulk and fitness, which not only produced a lot of heat but allowed him to keep going through the night. Above all, he too had sheer determination.

Ironically, many miraculous escapes from cold water occur for precisely opposite reasons. Young children, small and thin, and unlikely to struggle for long in the water, cool so rapidly that hypothermia can protect them. Every year, there are dozens of reports of children successfully resuscitated after being dragged from pools, ponds, rivers and seas after tens of minutes, or even an hour, of complete immersion without breathing. The exact mechanism of such miracles is unclear, but it seems to be down to a combination of the diving reflex with the fact that a cold brain has such a slow metabolic turnover, oxygen demands are virtually nil.

Seals and dolphins, while diving for very prolonged periods, slow their heart rate right down, and we humans seem to have a similar, perhaps vestigial, response to immersion. In our case, the reflex is triggered by water on the face, and if that water is very cold, our heart beat may slow profoundly. A small child falling into cold water may drop his or her pulse rate to just a few beats a minute, and this will limit the extraction of oxygen from the blood. At the same time, the cold will seep through the small body very quickly and within only minutes all the tissues, including the brain, will have cooled markedly. Their oxygen demands are then so tiny that the children enter a state of suspended animation, and the smaller you are, and the colder the water, the more likely you are to be revivable.

This message should never be forgotten. If anyone, particularly a child, is recovered from cold water, resuscitation should not be abandoned even if it appears to be hopeless. No reliance can be placed upon feeling a pulse, hearing the heart, or recognising breathing, for it is impossible to pronounce death with confidence while still out in the field. Recovery may yet occur if the victim can be rewarmed. An old adage among the cold experts has been proved time and again – nobody is dead until they are warm and dead.

*

Living on islands off the coast of Japan are the Ama people who have harvested kelp and pearls from deep cold waters for centuries. It is the young women who do the diving – demonstrating not only a remarkable capacity to swim vigorously under water while holding their breath for more than two minutes, but a capacity to resist hypothermia while they do so. The waters are often at just 7°C, and the women dive repeatedly through long working days, traditionally wearing nothing but thin cotton garments. Their apparent ability to withstand such cold stress has given rise to a number of studies of their metabolism, aiming to answer one question. Do they show an ability to adapt to the cold in the same sense that we all show adaptive changes to the heat?

Theoretically, humans might acclimatise to cold at the short-term, individual level or over evolutionary timescales at a population level. In either case, the adaptation would need to diminish heat losses or increase heat production so that a smaller fall in body temperature occurs with cold exposure. Studies seeking evidence for such cold adaptation, however, have generally found the opposite to be true. Men and women repeatedly exposed to cold, or originating from cold regions, tend to cool more easily than normal individuals. They are tolerant rather than resistant to cold.

Such cold tolerance is known as habituation rather than acclimatisation, and it can be very striking. The Australian Aborigines, Kalahari Bushmen and the Arctic Indians can all sleep outside with little or no clothing, even if their core temperature drops to less than 35°C. At such temperatures, you or I would be shaking too violently to rest and at the same time would be burning valuable food. A capacity to let your body cool can therefore be seen as a useful survival strategy, for not only do you get some sleep but you lose less weight while doing so. Nevertheless, such habituation does not appear to be a true evolutionary adaptation. The aboriginals' capacity to tolerate cold waxes and wanes with the seasons, and in the summer they shiver as easily as centrally heated Caucasians. Furthermore, Caucasians, if deliberately exposed to repeated cold, develop the same degree of tolerance.

The Innuit Eskimos of northern Canada also show cold habituation with no convincing evidence of true metabolic changes when compared to men or women from warmer climes. Nevertheless, some authorities suggest that they show a different type of adaptive change in that they are short and rotund, which reduces the ratio of their surface area to their body mass and heat-generation. This would make evolutionary sense, but not everyone is convinced. Growing up in a cold climate can also influence body development directly. Young pigs, for instance, develop short legs when raised in cold rooms, due to less blood circulating to their limb growing points. A similar effect may underlie the stature of the Innuit while, conversely, warm rooms following the spread of central heating may partly explain the increasing height of men and women from developed countries. In both cases, however, nutrition must play a part.

Returning to the Ama, we find the only convincing piece of evidence that genuine cold adaptation does occur. During the winter, the women divers not only become habituated to cold, but also demonstrate improved resistance to cooling. In part, this follows their becoming fatter at that time and so improving their insulation, but this seasonal change in body fat is seen all over the world, even in people who have central heating in their homes, and probably reflects lower activity levels in winter rather than climatic adaptation. In the case of the Ama, however, there is also a more marked shut down of blood vessels to the surface so that the flow to the skin and even the muscles immediately beneath is limited. This further reduces heat loss, and appears to be an example of genuine, short-term, physiological acclimatisation to cold.

*

I cannot leave the subject of cold without some discussion of hands and feet which, in many ways, are more of a problem than general hypothermia. During our Polar expeditions, Ranulph Fiennes and I have both suffered from frostbite in fingers and toes which have large surface areas for volume, low heat production, and are, quite literally, out on a limb. The blood they receive has therefore been cooled by travel down the arm or leg, but physiological defence mechanisms make matters even worse.

Shutting down blood flow to the skin and periphery, such as fingers and toes, is a good thing for core temperature protection in a true survival situation. Yet it is quite the reverse when it comes to coping with life and work in cold regions. The level at which evolution has set our hypothermic defence mechanisms is quite wrong. It is fine to sacrifice function of the hands and feet when a matter of life or death, but close down blood flow at the slightest hint of cold and the chances of getting into danger are increased.

The problem is that the temperature levels for cutting peripheral blood flow were set on the plains of Africa. Out on the savannah, the worst exposure to cold would have been at night, and then never extreme. Our sleeping ancestors would have benefited from reduced peripheral blood flow which retained heat in the centre of their bodies so that they could rest easily and burn less food. Cold hands and feet in the morning would have been a small price to pay. But following our late move into temperate and cold regions, the price escalated. Painful, sometimes frozen hands and feet are now a reality in extreme climates for we are stuck with a climatic protection system suitable for Africa, and entirely unsuited for the Poles or even northern Europe or the USA. Our hands and feet get cold in conditions which are otherwise quite comfortable and certainly not threatening, and this can be a major liability. Long before they actually freeze, they become too cold to be useful. With no dexterity, it is difficult to perform any tasks, and I have been caught out several times in situations where I have been unable to put on warmer clothing quickly because my fingers have already stopped working. Our cold defences literally threaten our lives.

There is, however, one brighter point. Unlike our very limited ability to adapt our cold defences in general, repeated exposure to peripheral cold does improve the situation with our hands and feet. The change involves a mechanism known as cold-induced vaso-dilatation, which operates as blood vessels in a finger or a toe drop to a stiffeningly low temperature of 12°C. Then they tend to dilate intermittently so that the limb is flushed with warmth. It is the explanation for that sudden and sometimes painful heating-up of fingers after twenty minutes or so of throwing snowballs with poor gloves on.

When first exposed to cold, the flushes of blood to fingers and toes are too infrequent to make much real difference, but after exposure for a few weeks marked improvements occur. Flushes come more often and for longer and overall fingers are warmer and work better. The adaptation is seen in all cold-dwelling races such as the Innuit and Lapps, but equally good responses are seen in the hands of fishermen, fish filleters and even postmen who work in cold climates. Its occurrence does not support the idea that any races have spent enough time in cold climates to have improved their physiological defences to the cold.

Compared with our fantastic abilities to adapt to the heat, exposure to cold can only induce meagre increases in heat production from shivering and a limited improvement in our insulation and the blood supply to our hands and feet. This, coupled with the hopelessly incorrect setting at which the blood flow to our periphery becomes dangerous, fits the theory that our ancestors left Africa just 100,000 years ago, that their progress to true northerly climes was later still and that evolutionary change runs slowly. The reality is that we remain adapted for sun worship, and it is only our clever behaviour that sees us through the heart of darkness.

EIGHT

★

Cries of the Heart

THE CLASSIC heart attack is unmistakeable. Out of the blue comes crushing chest pain, often followed by death. If the victim does survive, further attacks are likely in the next year or two and, even with no other episodes, there is a fair chance that the heart can no longer act as an efficient pump. Yet, despite the distinctiveness of such symptoms – recognised instantly today by anyone whatever his or her medical background – there were very few descriptions of such events in the medical literature before this century. Furthermore, the early physicians failed to write much about angina, the typical band-like chest pains that come on with exertion and frequently serve as a warning of a full blown collapse. The term was coined by the famous eighteenth-century physician William Heberden, but references to it are scarce. This is extraordinary. In centuries past, doctors may have been limited to cupping, bleeding and leeches as their therapeutic options, but they were generally assiduous and accurate recorders of disease. As late as 1892, Sir William Osler still described coronary artery disease as being 'relatively rare' yet less than two decades later, one in eight of all deaths in advanced nations could be attributed to it. By the 1980s, this figure had risen to a third of all deaths, and for every two people dying from a heart attack, three others suffered one. The increase in incidence has been terrifyingly swift. Coronary heart disease, unusual before 1900, has today become the primary peril, the biggest single killer in the western world.

What changed to make heart disease such a serious problem? The underlying cause of a heart attack is that the blood vessels supplying the muscles of the heart become clogged and narrowed. The blockages are due to inflamed fatty deposits forming both on the inside and within the walls of the arteries through a process known as atherosclerosis. Once narrowed, they cannot maintain a sufficient supply of blood to the heart, especially if it needs to work harder to meet the pumping demands of exercise. Even modest walking may cause cramping anginal pain, but this is just a prelude. When the vessels become narrow enough, a clot can form to block the artery completely. A whole area of muscle wall will then lose its blood supply, causing both intense pain and great danger. If the area of damage is extensive enough to stop the pump working properly, or it interferes with the electrical control of the heartbeat, death may follow swiftly.

If we examine what underlies the clogging process, a prime cause would seem to be smoking. The recurring theme of this book is that we no longer live in harmony with our evolutionary design and there can be no better example of this than drawing smoke, laden with chemicals and particulate irritants, into our delicate lung linings. It is clearly so unnatural and likely to cause harm that it is hard to understand how smoking was ever accepted as a reasonable thing to do. Yet, while nobody believes the contorted arguments of the tobacco industry that try to dismiss the link between smoking and damage to the lungs in the form of bronchitis and cancer, many people do not realise that the danger from cigarette smoke does not stop there. By mechanisms not entirely clear, smoking markedly contributes to atherosclerosis in the coronary arteries, as well as the furring up of blood vessels elsewhere. Many smokers will therefore have heart attacks and strokes, and may even lose their limbs, well before their time. As a practising hospital doctor, I can state without prejudice that I almost never see a patient under the age of 40 who has had a heart attack but who is not an avid proponent of the weed. For that matter, most of those I see in their 50s and 60s are also keenly addicted.

Despite the association between smoking and heart disease being well documented, tobacco has been around for much longer than the last hundred years. It is true that there has been much greater availability of cigarettes in the twentieth century, and no doubt this, along with the continued immoral advertising, has encouraged more people to smoke enthusiastically. In many Third World countries smoking is now more prevalent than ever it was in Europe or the United States, but in the less developed nations it tends to kill more people via the lung route, and so it is difficult to blame tobacco for all today's heart ills. There must be other factors that either contribute to the damage or else protect the hearts of Third World smokers from the ravages of cigarettes. An obvious place to look is in the diet.

Some decades ago, it was recognised that high circulating levels of cholesterol make it much more likely that you will suffer from coronary heart disease. Indeed, the relationship between your blood cholesterol and your vulnerability to a heart attack is almost linear. This led to the unpopular yet powerful advice that cholesterol should be avoided in the diet, and through the 70s and 80s health-conscious individuals were busy avoiding butter, eggs and cheese. Nowadays this approach is seen as rather too simplistic. Your body itself manufactures most of the cholesterol in your circulation, and so high cholesterol levels are more often a reflection of abnormalities in your metabolic handling of many types of fat rather than the direct result of eating too much food containing cholesterol. It is still a good idea to go easy on the butter, eggs and cheese, but the restriction is less rigid and the reasons underlying the recommendation have changed. It is not specifically cholesterol you should seek to reduce but the intake of fats in general.

Cholesterol is carried around the body linked to specialised proteins. Essentially there are two forms – high density lipoproteins (HDL) and low density lipoproteins (LDL). Taking a rather simplified view, cholesterol in HDL is being carried to areas where it is needed, while that in LDL is the excess which is to be put into storage. People with high levels of HDL usually keep their blood vessels clear, while those with high levels of LDL seem inadvertently to deposit fats in their arteries. The difference is important. It is not the total level of cholesterol in your blood that matters but the ratio of HDL to LDL: high and you are relatively safe, low and you are in danger.

The ratio of HDL to LDL is little influenced by the intake of high cholesterol foods alone, being more related to excessive intake of other types of fat. Our western diet contains many different fats and oils, and it has only recently been appreciated that, far from being merely a ready energy store, their metabolism is exceedingly complex. Too much fat is definitely harmful, but the balance of different kinds in our diet seems more important than the overall quantities consumed. In recent years research has focussed on three areas: the potential dangers of saturated fats, coming mainly from animal sources; the possible protective effects of consuming specialised unsaturated fats, such as those found in fish and some vegetable oils; and the body's handling of the hydrogenated fats, which come in a form that has been chemically altered by man. Each of these can be examined from an evolutionary perspective, bearing in mind the epidemic of chest pain that started early in this century.

It is not in doubt that a high intake of saturated fat can increase LDL cholesterol levels in the blood. Your consumption of saturates is largely dependant on the amount of animal products in your diet in the form of both meat and dairy produce. The fact that such products are potentially harmful has encouraged some enthusiasts to claim that humans must have been designed to be vegetarian. It is clear from the mechanisms of our teeth and gut, as well as from our metabolic need for certain vitamins, that this cannot be true. Our ancestors must have been omnivorous for several million years, although the balance of their diet was probably quite different from our own.

In contrast to real carnivores such as the leopard, our predecessors were certainly no great hunters. Their ability to find and kill prey would have been, to say the least, inefficient. We must therefore have eaten a diet which, by modern standards, would have been towards the vegetarian end of the spectrum, and this would have changed little until the development of agriculture brought greater access to meat and dairy products. The latter, in particular, are interesting. Today the eating of foods derived from milk is often thought of as a return to the natural diets of our past, but before the domestication of cattle and goats our chances of getting much milk would have been remote. The butter and cheese that now contribute a lot to many people's diet is even less in line with our evolutionary design than a high meat intake, and a substantial proportion of the world's population lacks the enzyme in the gut to allow proper digestion of cow's milk – a fact overlooked by vegetarians advocating their practice for 'return to nature' health reasons rather than for those of animal cruelty.

On the other hand, as with smoking, if a high intake of animal fats were the prime cause of atherosclerosis, why have heart attacks become so much more common only in this century? The consumption of meat and dairy products has been widespread for many thousands of years, and although better farming methods and increasing affluence have probably led to greater intakes by more of the population, those who would have eaten a lot in centuries past would also have been those seen by the doctors. Physicians have always been the prerogative of the rich. So, while animal fats may contribute to heart disease, it is difficult to point the finger at these alone.

In recent years medical science has confirmed that consuming fats from oily fish grants some degree of protection. This was first suggested by observations on the Innuit Eskimos of northern Canada, who traditionally have eaten a diet that is extremely high in fat but who exhibit a remarkably low incidence of heart disease. Their consumption of fish and marine mammals in place of conventional meat and dairy products appears to be the likely explanation, although the Innuit might also have a genetically determined resistance to atherosclerosis.

To examine these possibilities, studies have been made of other fish eating communities. The Japanese are great consumers of sea food and have a low incidence of heart problems when dwelling in their native land. Those who migrated to the United States in the 50s and 60s and enthusiastically adopted American eating habits are not so fortunate. In fact, they face the same risk as their fellow Americans of suffering from heart complaints, and that is many times the risk of their cousins back home. Since they have refrained from much intermarriage with other U.S. citizens, the genetic make-up of the migrant groups has changed very little, and so their consumption of meat in place of fish appears to be the key. Indeed, it has been shown in Caucasian populations that those who eat a portion of fish a couple of times a week suffer from heart disease less frequently.

Of course, the health benefits of higher fish intake could just be a consequence of eating less animal fat rather than a positive effect of consuming fats of marine origin, but this is unlikely. Taking fish oil in the form of capsules will reduce your chances of suffering a further heart attack after surviving a first, even if you make no other changes to your diet. So some of the benefits must be direct, and there are a number of ways in which these may occur.

The membranes of all our cells are constructed from specialised fats built from components that are taken in through our diet. Marine oils alter the properties of these membranes, including those of small cell fragments in the blood called platelets. These are designed to adhere to breaks or damage in blood vessels before they trigger the system that builds a fibrous clot, and the incorporation of marine oils in their membranes significantly reduces the stickiness of the platelet surfaces. Since heart attacks are caused by clots forming at narrowings in the coronary arteries, less sticky platelets make it less likely that a clot will block things off.

Substances derived from fats are also involved in the body's inflammatory system which responds to trauma, infection, cancer and so on. The system is in careful balance, with some elements encouraging inflammation in order to ensure that protective white cells get to the areas of damage to make repairs, while others are designed to control the extent of the process and to damp it down again when it is no longer needed. The equilibrium seems to be influenced by dietary fat, with animal fats tending to increase inflammation, while marine fats help the anti-inflammatory side. Although inflammation is probably not the primary cause of the deposition of fat in blood vessel walls, once formed the deposits themselves become inflamed, and this can make them grow larger. The effects of marine fats on lowering the incidence of heart disease may therefore be dependant on their reducing inflammation.

A teleological view would suggest that if marine oils in the diet are beneficial to health, evolution must have set us up to work with them, which in turn indicates that we must have spent some time in our evolutionary development exposed to a diet high in sea foods. I have mentioned previously that some anthropologists believe our divergence from the chimps was triggered by our isolation on an island where we lived as semi-marine apes. At first sight, this would seem to fit the bill, explaining why fish oil became so good for us, but since that divergence took place just five to seven million years ago, and Australopithecines were wide-ranging by four million years ago, it is difficult to place our ancestors in a littoral setting for more than two or three million years, even if the theory is true. Although this was clearly a long enough period to effect a change in our posture, it was so because becoming erect conferred very considerable survival advantages precipitated by the move from the forests to a different environment – whether the seashore of the marine theorists or savannah of conventional thinking. It seems inconceivable, however, that so short an evolutionary period would alter our inflammatory processes for they are fundamental to mammalian biology. In fact, the way that inflammation works is virtually identical throughout the higher animal kingdom, and so it seems much more likely that the benefits of fish oils are a remnant of our truly distant marine origins, before the evolution of mammals. Yet, whatever the history, a change in our intake of fish could not account for the relatively recent increase in heart disease anyway. Many may have eaten fish every Friday before 1900, but very few of the population would have done so more regularly.

The introduction of hydrogenated fats with an artificial chemical structure at the beginning of this century certainly did coincide with the rise in heart disease. The hydrogenation process allows liquid vegetable oils to be made into solids or semi-solids by chemically altering some of their bonds. Margarine and other related fats began to appear in many of our foodstuffs and have since been used widely in the baking of biscuits, bread, cakes, and pies. Hydrogenated fats contain structures that have never been seen in nature, and we may be poorly adapted for them. The way in which the body handles these unnatural chemicals is only now beginning to be understood, and they certainly have an adverse effect on the ratio of HDL to LDL. Most experts believe, however, that their influence is small, especially as there is mounting evidence to suggest that we should also focus on many other areas where we have deviated from our natural behaviour and food intake.

Vitamins and trace elements in fruit and vegetables, particularly the anti-oxidants, are now thought to underlie the health benefits of the so-called Mediterranean diet, and these include a reduction in heart attacks. Vitamin C, E and Beta Carotene are just some of these anti-oxidants and all are more common in diets with a high vegetable content. Oxidation damage is caused by the metabolic production of free radicals – reactive atoms or ions – which rapidly combine with the nearest proteins or membranes. It is nature's equivalent of rusting. Because free radicals are so harmful, evolution has developed a complex system of defences against them, largely dependant upon substances which can combine with them before they have a chance to damage body tissues.

In dietary research on 700 people over the age of 65 carried out in Britain in 1973, it was found that the chances of their dying from a stroke in the next ten years were highly correlated with their vitamin C intake from citrus fruits and vegetables. Those who ate a lot were less than half as likely to have a stroke as those that ate a little. It is difficult to be sure that vitamin C was solely responsible since diets high in one anti-oxidant often contain high levels of other vitamins as well. An alternative explanation, not recognised at the time, is that the benefits of diets rich in micronutrients may be linked to the metabolism of a substance called homocysteine.

Homocysteine is an amino acid, both consumed in our diet and made by the liver. Reasonable amounts are needed to build our proteins, but it is now known that excess quantities contribute in some way to the furring up of our arteries. Recently it has been shown that many people have levels high enough to be a danger, not due to eating more than is healthy but because the liver makes too much. Excessive homocysteine production in the liver results from a group of chemical pathways not working properly due to an inadequate supply of one or more of the vitamins B6, B12 or folate. In the past, a lack of these vitamins was thought to be significant only if sufferers developed sore mouths, bad skin or anaemia. Now it is clear that long before this level of vitamin deficit is reached, liver malfunction can disturb the balance of amino acids in the circulation. Today's evidence suggests that one third of the 30-year-old population already has a high homocysteine level, while among the elderly the proportion rises to two thirds. It is encouraging to know, however, that even the elderly can reverse the abnormality if given the vitamins as supplements, although this does not mean that we should all start popping pills. Eating a healthy and mixed diet is a sounder approach.

Could then a change in vegetable intake in this century have been the predominant cause of the increase in heart disease? I think not. Vegetable consumption has certainly diminished in the developed world, and this is cause for concern, but the decline began after the Second World War rather than earlier, and so it would be a mistake to ascribe the majority of our problems to that change.

Another dietary factor that has received much attention recently is our intake of salt. Back on the African plains, salt losses in the heat would have been significant despite acclimatisation limiting the concentration of the salt lost in sweat and urine. In those conditions we did well to eat as much as we could get, and so evolution gave us a taste for salty things. Now in the era of snacks and canning, salt has become a significant problem. Men and women in modern societies eat up to one hundred times more salt than our early ancestors, and for some decades it has been thought that this excessive intake might contribute to high blood pressure. Recently this has been confirmed by the 'Intersalt' study which compared blood pressures in more than ten thousand people in 32 different countries with their salt intake. The results were irrefutable; the more salt people ate, the more likely they were to suffer high blood pressure, and the effect was more marked with age.

One of the difficulties in treating hypertension is that, unless it is very severe, when it may cause headaches or illness from kidney damage, it usually displays no symptoms. There may therefore be long delays in diagnosis and it is sometimes hard to persuade individuals to accept treatment when they feel nothing wrong. Medication can also have side effects. Nevertheless, it has to be taken seriously. A high pressure in the arteries leads to thickened walls which are more prone to internal damage and the fatty deposition of atherosclerosis. Men and women with high blood pressure are at much greater risk of heart attacks and strokes. To make matters worse, the blood vessels that supply the kidneys are particularly prone to atherosclerotic damage and normal blood pressure is largely under renal control. When the kidneys sense that blood pressure is low, they release hormones in order to raise it so that more blood will be received for filtering and cleaning. This is fine while the pressure they are sensing is that of the whole body and not a lower pressure caused by local narrowing of the vessels that feed the kidneys themselves. But often the kidneys are misled into releasing hormones to raise pressure when it is already high, and then a vicious cycle ensues. In the end, levels can become so high that strokes, heart attacks and kidney failure are almost inevitable.

Unfortunately, throwing away the salt cellar would make little difference. Our taste for salt has not gone unnoticed by those employed in the food production industry who know that the more salt they put into their products the more their customers will like them. Salt content is often high where least expected. There may be more salt in a bowl of cereal than in a packet of crisps. Furthermore, salt in foods such as sausages is associated with a higher water content, making them heavier and effectively cheaper to produce. The net result is that in Britain, one in four adults over the age of 45 and one in three over the age of 60 are hypertensive; the cost to the National Health Service in drug treatments is some £300 million pounds a year. Industry knows this but has never been keen to put health before profits. The only solution is to use more fresh produce and to cook and eat without adding salt. If you do this, you can at least take encouragement from the fact that the foods will not taste bland for long, and soon it will be the salt-rich foods of industry that taste nasty.

Overall, when it comes to our diet, there are clearly many aspects which are out of step with our biology, and these contribute to our propensity to develop heart disease. Nevertheless, taking account of the changes in our diet between this century and the last, it is difficult to identify a single reason for the increase in heart attacks. We are therefore left with an incomplete explanation for the problem and must look elsewhere to find the cause of our ills.

*

I discussed earlier in this book why, in the distant times of the hunter-gatherers, all men, women and children would have been active for most of their lives and, although less marked, active lifestyles must have continued for the majority right up to this last century. Even if occupation did not entail much action, most people would have walked from place to place. Since the Industrial Revolution, however, the development of urban conurbations, with housing close to places of work, and the invention of motorised public and then private transport has seen this progressively change. Are low activity levels at the root of our heart problems?

In the Allied Dunbar health survey of Britain carried out in 1994, adults were rated on a scale that assessed the number of episodes of moderate to vigorous exercise undertaken in a typical week. The results showed that for middle-aged groups between 45 and 64 years old, those who did no exercise at all had heart problems in 16 per cent of males and 11 per cent of females, while the figures dropped to just 4 per cent and 2 per cent of the same age who exercised at least five times a week. Even exercising just once or twice made a considerable difference. Such observations are striking, and it is reasonable to assume that, before this century, most adults would have been getting enough exercise to qualify for the five times per week groups. A change in activity levels can therefore provide an explanation for about a four-fold increase in the incidence of heart disease.

Exercise can influence the heart in many ways, primarily making it larger and stronger in order to meet demands for it to pump blood faster. With the increased thickness and strength of heart muscle walls, the blood supply vessels also enlarge so that any furring within them has less of a narrowing effect. At the same time, the expansion prompts the development of new blood vessels which make any one region of heart wall less dependant upon the supply from a single artery. Further benefits are gained from a change in platelet stickiness. Just as the consumption of fish oils reduces the likelihood of platelets adhering to areas of damaged artery wall and generating clots, regular exercise has the same effect. An experiment with which I was involved while working for the Ministry of Defence illustrated this most strikingly.

At that time, I was advising the Army on the types of food that would best sustain soldiers in the field. There is a difficult balance to be struck here, for troops must often carry their food on their backs, and so eating more to feel better and stronger has to be offset against the increased weight of their loads. Using high fat rations would allow more energy to be carried for the same weight, and over the likely period of active operations should not be detrimental to long-term health. But sometimes soldiers must work at high intensities, the sort of levels at which muscles prefer carbohydrate for fuel. My research examined whether using high fat rations to save weight might reduce their working capacity.

To answer the question, we took two comparable groups of soldiers and made them compete against each other in performing hard military exercises over bogs, moors and fells for many hours every day. The training continued for a whole week, and when they arrived back at camp each evening, we took them into a temporary laboratory and made them work on exercise bikes until they were so utterly exhausted that they could do no more. The time that they could keep going on the bikes in the evenings was used as a measure of their endurance reserve at the end of each day's efforts.

During the biking, we took blood samples in order to monitor the soldiers' metabolism. Little plastic cannulae (the type used to give drips in hospital) were inserted into the veins in their arms so that small samples could be taken every fifteen minutes without stopping their work on the bikes. On the first evening of the study, I found in ten of the sixteen subjects the cannula blocked off by a clot which had formed between samples. By the evening of the third day, the number blocked was down to three, and for the last couple of days there were no blockages at all. Quite apart from the main purpose of the study, the subjects had shown that a vigorous exercise programme completely changes the tendency for blood to clot and block off narrow tubes. It was clear evidence as to why an active individual may not have a heart attack even if he or she has diseased and narrowed coronaries. A more important benefit, however, is that regular exercise makes athero-sclerotic narrowings far less likely anyway.

Muscles burn fat at all but the highest intensities of work, and regular activity modifies the fat profile of your blood and eliminates a good part of any excess intake. Furthermore, when endurance trained, the muscles have enlarged fat stores within them, ready for use as an immediate source of fuel. As part of this adaptation, the HDL transport system becomes more efficient in order to carry fat to the muscles where it will be needed. Regular exercise will therefore produce a healthy ratio of HDL to LDL and so reduce the risk of fat being deposited in the wrong place. Such benefits are evident even on lazy days. Studies of volunteers who ate a fatty meal on the day following a bout of vigorous work showed that fat levels in the blood an hour after the meal were 30 per cent lower than when eating the same food on the day following no such activity.

A similar but more extreme example of the effects of exercise on fat metabolism were to be seen in the blood samples taken from Ranulph Fiennes and myself during our Antarctic crossing. As explained earlier, we ate a very high fat diet during our walk in order to minimise the weight of the sledges we dragged, and the fat source used was mainly butter. Our food was not only high in total fat but very high in saturates. The World Health Organization recommends that you eat a diet containing no more than 30 to 35 per cent fat, of which the larger proportion should be unsaturated. Our Antarctic rations contained 57 per cent fat, and that was as a proportion of a diet containing more than twice the normal adult consumption. The result was an intake of more than four times the saturated fat normally considered wise, yet our total cholesterol stayed at healthy levels and the good HDL type went up while the bad LDL went down. The message was clear. If you exercise enough, the fat gets used for the purpose that nature intended, and you can eat the most awful of diets and get away with it. And there is another reason why this is so. Since exercise burns calories, people who undertake it tend to eat more food. This gives them more micronutrients, which in turn can protect their health.

It is a serious matter for the medical profession as well as the public to find two thirds of the elderly deficient in vitamins to the extent that they develop unhealthy homocysteine levels. In some cases, the shortages can probably be ascribed to factors such as illness, poverty and social isolation, which lead them to eat poor quality food and a diet lacking fresh vitamin-rich produce. Yet, on the whole, the elderly adhere to a diet which would contain plenty of micronutrients if they ate anywhere near enough of it. The problem is that their micronutrient requirements are very similar to younger more active adults but their sedentary lifestyles demand little food to meet their energy needs. Evolution simply did not expect anyone, at whatever age, to stay still for so much of the time. It set us up to extract vitamins from a high throughput of food. My fear is that the problem will get much worse. Today the older generation tends to eat a healthier diet than the young, with more fish, vegetables and fruits. If they cannot extract enough vitamins from that type of food, how will the younger generations fare as they grow older, consuming packaged foods of low micro-nutrient content and exhibiting a tendency to exercise even less?

In conclusion, optimal health is to be gained only through attention to both food and exercise. If we want to avoid the cries of the heart, there is no doubt that we should never smoke, and we should reduce our intake of fats from animal sources and eat more fish, fruit and vegetables. But along with these changes, an increase in activity is needed. This is an uncomfortable fact for those who currently do little, and many people will be unable to alter their lifestyle to counter what they see as a theoretical risk of that chest pain somewhere in the distant future. However, for such people there is perhaps a more powerful motivating factor close at hand. You are unlikely to become obese or even unfashionably overweight if you exercise regularly.

NINE

★

Survival of the Fattest

EVEN MORE than the epidemic of heart disease, our propensity to get fat reflects a biology out of step with our lifestyle. Wild animals, including all of our close relatives among the primates, are virtually never plump, and the same must have been true of our ancestors through the whole of human evolution. Now more than half the adults in Britain are overweight and one in seven are obese, not just too heavy according to some whim of fashion but fat enough seriously to endanger their health.

Recently, some groups have been trying to promote the idea that being overweight is really nothing to worry about. Particularly vociferous in the United States – where one in four adults are clinically obese – has been the punchily named National Association for the Advancement of Fat Acceptance. The members of this organisation proudly claim that fat is both good and beautiful, suggesting that the only problem is an unfair dislike of fat people. There is some truth in their highlighting this prejudice. The social slur of being too fat is easy to dismiss as exaggeration, but the consequences can be cruel. It is particularly harsh for overweight young women, who complete fewer years at school, marry less frequently, end up with fewer friends, and have more episodes of significant depression than their lighter equivalents. However, while rating fattism as bad as racism, I also believe that we should avoid promoting the message that becoming overweight is all right. In reality, this is very far from the truth.

It has been estimated that diseases caused by obesity cost the National Health Service more than two billion pounds a year, and that a further similar sum is spent on the largely ineffective slimming industry. It is also reckoned that Britain loses up to fifty million working days per annum due to ailments related to obesity. The cost to industry amounts to a further £2.5 billion. These are very significant sums, yet it is not the economic cost that is the real problem. Being too heavy produces all sorts of medical problems and, ultimately, is life-threatening.

The most striking additional risk to health for fat people is their greater tendency to fur up the arteries to their hearts, brain and other organs. Being over ideal weight by just 3 kilogrammes (6.5 pounds) makes a measurable impact on the chances of your having a heart attack. If you have reached the point where you are classified as clinically obese, your vulnerability is more than doubled. Many of the factors damaging your blood vessels will be the same as those causing atherosclerosis in normal weight men and women, but occurring more frequently and severely. The overweight frequently have high blood pressure and adverse ratios of HDL to LDL accompanied by abnormally high levels of other unwanted lipids in the blood. Fat abnormalities in the blood are partly explained by the simple fact that the obese eat far more than their thin counterparts. The idea that they have a low metabolic rate and became overweight despite eating little is a myth. Those who are heavy tend to be less active than those who are light and so gain few of the benefits of exercise. They also develop abnormal metabolism of nutrients and this may effect the body's handling of sugars even more than disturbing levels of fat. Indeed, many of the overweight become diabetic as they grow older.

The diabetes of obesity is not usually the form that occurs in children or young adults. That type follows the destruction of cells that produce insulin in the pancreas by an inflammatory process with an unknown trigger. It is usually very severe as soon as it starts and almost inevitably requires long-term insulin for treatment. The overweight, on the other hand, develop a much less dramatic form, coming on slowly in middle or older age. It is due to the pancreas being unable to cope with a lifetime of high demands from an oversized body, combined to some extent with a propensity to eat too much sugary food. Unlike complex carbohydrates, such as starch, which would have been important components of our diet for millions of years, simple sugars are in relatively short supply in natural habitats. Our sweet tooth – an evolutionary development to help us identify ripe fruits – did not lead us into trouble until refined sugar became widely available. Analogous to salt, the situation was then worsened when food manufacturers started using it indiscriminantly without considering whether it might be unhelpful to health. The amount of sugar in a can of baked beans is a prime example of exploiting an evolutionary weakness for profit.

To be fair, not all 'age-onset' diabetics are overweight, but it has been estimated that more than 70 per cent of older diabetics would not have developed their illness if they had not been too heavy. Fortunately, many cases can be treated by diet and weight loss alone, although it does require considerable determination. For those who do not have it, control of sugar levels by tablets or insulin injections is usually highly successful, although excessive sugars are only one element of the metabolic problem.

In most diabetics, the abnormalities in the body's handling of fat are almost as great as those of blood sugar control. Frequently they also have too much circulating cholesterol, with markedly adverse HDL to LDL ratios, and the difficulties these create are not easy to treat. Even on medication, diabetic patients remain prone to developing athero-sclerosis, and few manage to reduce the risk by controlling their diet and weight or by taking up exercise. Years of being overweight, complicated by the diabetes it has caused, then leads to angina, heart attacks, strokes, aneurysms and kidney damage – all dangerous medical conditions. Furthermore, the sheer load imposed by a large body, with or without diabetic complications, can simply overwhelm the heart. With a circulation too sluggish to clear fluid from lungs or limbs, breathlessness and swollen ankles may herald an early death.

Obesity may threaten health in more ways than just metabolic derangement. Snoring is usually thought of as harmless. It occurs when the muscles of the throat relax in sleep, obstructing the flow of air. A heavy neck makes it much more likely, and fat people snore more often and more loudly. This is not only inconvenient for partners; heavy snoring carries health risks beyond the dangers of a sleepless, irritated spouse losing self-control. In severe cases, almost always in the fat, the airway can obstruct completely and prevent air flow. Often this occurs in cycles through the night, when louder and louder snoring culminates in a complete cessation of breathing that can last up to half a minute. The snorer then wakes, fighting for breath and, without becoming fully alert, their sleep is disrupted. Then they lapse back into somnolence and repeat the cycle once again. It is more than just annoying. Nights become so disturbed for snorers that they are sleepy during the day and prone to accidents. Furthermore, periods of not breathing at all cause low oxygen levels in the blood which slowly damage heart and lungs.

Apart from this, what of the impact of carrying all that extra weight around? Walking with only a light backpack considerably influences any task and, for most of us, going up hills or stairs is considerably more taxing when carrying just 5 or 10 kilogrammes (10 to 20 pounds). Yet the obese may carry an extra 30 kilos around with them all of the time. To make matters worse, they also tend to overheat due to the insulation of the fat beneath their skin. The result is like working hard in winter clothing. Hot days – a delight for most of us – are no fun at all for the overweight and summers can become a nightmare of discomfort. Apart from the survival advantage when escaping from a sinking ship in a cold ocean, being too fat has nothing but downsides. At best, it interferes with most days of your life, at worst it will rob you of many years.

*

If being too fat is so harmful, we should have evolved regulatory systems to prevent it. Why then is it such a common problem, and what factors dictate why some people stay thin while others grow so large? The answers must be matters of balance. If the energy contained in your food exceeds what you need, you will gain weight; if it is less, you will lose it. This is an absolute truth, yet it has proved remarkably difficult to pin down all of the factors on either side of this energy balance equation. When I first developed an interest in weight regulation, I found this astonishing. Now I realise the difficulties inherent in accurately measuring all the factors that contribute to energy intake and output over long periods, and have come to appreciate that tiny discrepancies between them can lead to huge changes in weight over the long term.

In 1995, newspapers in the United States reported the death of the fattest man in the world. He had come to weigh 465 kilogrammes, or around 1,000 pounds (in excess of 70 stone). On a visit to hospital he had to be transported by fork lift truck, and after his death the wall of his bedroom had to be demolished in order to remove the body. Obviously he was an extreme example of the obesity problem, yet it would have required a weight gain of only 37 grammes per day to take him from a normal 70 kilogrammes (155 pounds) at the age of sixteen up to the grotesque proportions of his death at just 45 years of age. This is the equivalent of eating an extra 250 calories per day – less than one small bar of chocolate.

To complicate energy balance assessments, factors on both sides of the equation are liable to change. The alterations occur because evolution has developed weight regulatory systems and these constantly attempt to match intake and output to prevent either excessive weight loss or gain. Unfortunately, when it comes to the latter, some aspects of modern life seem to deceive our controls. Many find weight gain all too easy and then face considerable difficulties when it comes to reversing the trend. It is a sad fact, but once you are too fat, your body does its level best to keep you that way.

Weight control mechanisms can act on either intake or output, and in animals they influence both. On the intake side, animals are rarely exposed to prolonged periods of excess food because of adjustments in population. When food is plentiful, they become more fertile and are more successful at raising their young. The result is a rapid increase in the number of mouths to feed, although usually there is simultaneous growth in the numbers of competitors and predators. Among people living in the under-developed world, such population factors may still have some effect. A successful harvest can influence birth rates, since human fertility is also improved when we are well nourished. Infant survival is also better when food is plentiful, and a good supply permits earlier weaning, whereas times of shortage may see breast feeding, with its contraceptive effects, extended for several years. In the developed world, however, no such factors operate. People's breeding habits, the success of their pregnancies and the survival of their young are entirely independent of the number of ready meals in the fridge.

Intake control in animals is not just a matter of population adjustments. Even when there is a glut of food, individuals tend to eat only enough to match their needs. They must therefore have mechanisms to match appetite to requirements, but it is not altogether clear how these work. Day to day, the amount of food in their stomachs and the level of sugar in their blood are known to trigger signals which make the animal feel appropriately hungry or satiated. Beyond this short-term regulation, a background control also seems to operate, keeping body weight stable over weeks or months. Recent evidence suggests that it is seated in their fat stores.

Genetically obese mice have been shown to lack the normal form of a hormone produced by fat cells which suppresses appetite. It has been called 'leptin' – after the Greek *leptos*, meaning slender – and is released by fat stores when they are full. After the abnormality was identified, it was found that injecting normal leptin into these fat mice made them lose weight quickly. The discovery led to a frenzy among bio-technology companies and a down-payment of $20 million was made for the rights to exploit the gene. The hope is that an artificial human leptin will be God's gift to dieters, and the dream has been boosted by the recent discovery that human fat cells do carry a leptin gene. Development of the discovery will take a lot more investment, but countering human obesity is very big business.

From an evolutionary point of view, leptin must be part of a system that ensures fat stores are fully stocked whenever possible in order that either animals or humans can face the next period of famine. It makes survival sense, and although, as yet, there is little evidence that human leptin is effective in suppressing our appetite, most dieters recognise that as soon as they relax their self-imposed discipline, they are driven to eat more until they have regained all the weight they have lost. Leptin could well underlie this drive, and my own experience certainly lends credence to its power.

When Ran Fiennes and I returned from our Antarctic journey, body composition studies revealed that our fat stores were completely empty. The very hard work had led us to use far more energy than we had eaten and our weights were down by nearly 25 kilogrammes (55 pounds). With this weight loss, we were absolutely ravenous which, to a non-expert, might not seem surprising. Nevertheless, it was actually unexpected – most people who starve to such low weights are profoundly anorexic. Starvation probably causes deficiencies in essential micro-nutrients as well as deficits in energy, and a stack of fat, protein and carbohydrate is the last thing a weakened body needs if it lacks the nutrients required to process the food and build new tissues. Anorexia after starvation must therefore have evolved to protect us from refeeding dangerously quickly. Some micro-nutrient deficiencies, such as a low zinc level, even cause complete loss of taste in order to dissuade us further from over enthusiasm. Before these issues were understood, medical refeeding of famine victims sometimes did as much harm as good. Even today, some doctors thoughtfully provide artificial feeding for patients who have starved through illness, without attending to the micro-nutrient deficiencies first. They are then surprised when the patients unexpectedly die.

Ran and I were not like famine victims. We remained hungry in the face of great weight loss because, despite eating too little to meet our needs, we had maintained a very large throughput and so had no marked vitamin or trace element deficiencies. Indeed, rather than anorexia, our drive to consume was total and we ate both day and night for several weeks after coming home. This was well illustrated by a situation arising about a month after our return. Waiting one lunchtime to give an interview for BBC radio at a London studio, I found myself in one of the hospitality suites faced with a large plate of sandwiches. I proceeded to eat them all, and ended up both uncomfortable and embarrassed. The sandwiches were meant for the five other guests on the show as well as myself.

Following my interview, I was to speak on a different programme, to be recorded a short distance away. With plenty of time to get there, I decided to walk along Oxford Street in central London. Although I am no hamburger fan when it comes to health, I am none the less hypocritical enough to eat one occasionally. During the crossing of Antarctica I had dreamed of such an opportunity, and now I had my chance. I went straight into the first McDonalds I came to, despite already feeling full from lunch at the first studio, and ended up eating six hamburgers before I finally reached my destination – not all at the first McDonalds, but three at other branches of the same chain and two at Burger Kings. I simply could not walk by these hamburger joints without going in, although discomfort turning to pain did limit me slightly. Only the first four burgers came with fries.

What drove this feeding frenzy? Like animals, humans have short-term triggers for hunger, including those that report how full our stomachs feel and our levels of blood sugar. But often these natural triggers are overridden by cultural considerations along the lines of 'it's twelve-thirty and so it must be lunchtime'. In the case of my hamburger blow-out, none of these applied. My stomach was full before the first burger, my blood sugar levels must have been normal, and my recent sandwich meal should have allayed the 'time for lunch' excuse. Something else then was driving me to eat, and empty fat stores with low levels of leptin must be a likely cause. The notion is also supported by the fact that such dedicated piggery lasted for six weeks after the expedition before it switched off, almost overnight. That was the time it took my fat stores to return to normal, and Ranulph Fiennes behaved very similarly.

If leptin, or something like it, is responsible for controlling human appetite, it should suppress our eating when fat stores are full and so prevent them from becoming overloaded. What then goes wrong in so many men and women?

One suggestion is that fat people have defective visual cues for eating. In an experiment some years ago, students were invited to participate in a study which involved completing some questionnaires, but when they arrived at the study centre and were asked to wait their turn, they were each offered hospitality in the form of drinks and sandwiches placed on a table in front of them. They were also told that there were more sandwiches in a nearby fridge if they wished. Unknown to the students, the study had nothing to do with the questionnaires but instead examined the number of sandwiches they ate while waiting. They were brought back several times to the study centre at various intervals from their last meal, and there they faced different numbers of sandwiches on the table – sometimes too many, sometimes too few, but always with plenty in the fridge. The results were startling. The number of sandwiches that students of normal weight consumed varied according to when they had last eaten, and so corresponded to how hungry they felt. Fat individuals, on the other hand, showed no such regulation. They did not necessarily eat more, for they appeared reluctant to raid the fridge, but they did eat whatever food was on the table, regardless of how much it was or how hungry they felt. Their eyes appeared to rule their stomachs.

Although this phenomenon may contribute to men and women becoming overweight, I doubt that it lies at the root of the problem. There is strong evidence that the type of food we eat can confuse our intake regulation. Indeed, it can even override the intake regulation of animals. Laboratory rats, fed on standard chow, control their appetite very accurately and never get fat however much food is put in their cage. But if offered what is known as a 'cafeteria diet' – meals consisting of snacks such as hamburgers, sweets and chips – they completely lose this self control and can become positively enormous. Several reasons probably underlie this: the diet has an attractive mix of tastes and so encourages feeding; it has a high fat content so that more calories will fit into the same degree of stomach filling, and the fat, unlike protein and carbohydrate, can slip into the body fat stores with virtually no metabolic cost. Is the human propensity to become overweight analogous to that in 'cafeteria' fed rats?

Many studies have shown that if we eat a meal with a high fat content but the same number of calories as a more balanced dish, we perceive the high fat meal as less filling. This is not surprising. The high fat foods contain more calories for weight, and so the meal is physically smaller. They also contain less sugar, and as a result cause a smaller rise in blood glucose. Our ability to assess the calorie content of the higher fat meal is poor, and when our diets are rich in such foods, we will tend to eat more than we need. Nevertheless, this is only a part of the problem.

The fact that we derive particular pleasure from high fat foods is probably not just chance. Cream cakes or curries bathed in oil are far more popular than any insipid healthy equivalent. Fatty foods taste good. There is probably an evolutionary basis to this. The body can assimilate additional fat at almost no metabolic cost, for it is simply transported directly into stores for future use. By eating as much as possible when supplies were good, our ancestors would have been set up for the bad times that might lie ahead. Fat gains are also a good way to prepare for winter, when the cold can be countered by improved insulation. Evolution has therefore designed us to overeat whenever possible, although this does not mean that our forbears were fat. Over-eating arose as a problem only when the availability of food became almost unlimited and the famines no longer arrived. Today we build our survival stores but never put them to use. Civilisation has made our lives as unnatural as those of the caged laboratory animals.

Additional factors may also operate. Contained in our foods are those micro-nutrients essential for our well-being and little is known about how their consumption is regulated. The processes of natural selection, however, should have provided us with means whereby our bodies could identify shortages and take measures to correct them. These responses might be non-specific, driving us to eat more of any food when a vitamin is lacking, or targeted, making us want to eat more of the types of food rich in the missing nutrient. Either way, such responses could contribute to obesity. Most diets of today are less nutrient rich than those of our evolutionary past, and by cooking them we further diminish their vitamin content. Hamburgers and chips may make us overweight through two very different routes. On the one hand, the fat content appeals to our tastes and slips past our regulation system, while on the other, the food lacks so many of the things we need that we are driven to eat more in a bid to sustain our health. Perhaps this is best described as a double whammy.

*

As well as by intake controls, animals alter the energy they use to help maintain balance in the face of food excess or shortage. Some adjustments are simple: to gain more food in the natural world will take more effort, and if the animal does become heavier, it will need more energy to get around. Other adjustments are more complex. Animals can change their metabolic efficiency.

The amount of energy used by an animal or a person depends on three elements: their basal metabolic rate – the energy needed to maintain life when resting; thermogenic responses – increases in metabolism triggered by factors such as cold, food intake or fever; and the energy cost of activity. All three elements can be altered to maintain stable body weight.

When food is in short supply, animals lower their basal metabolism, partly because they lose body weight and metabolically active tissues, and partly because cell activity is damped down to conserve energy. When it comes to over-eating, similar but opposite changes take place. An increased body size uses more energy and metabolic activity is uprated. Other interesting alterations are to be seen in the animals' thermogenic responses, which include the metabolic reaction to cold and the increase in metabolism following a meal as it is absorbed, digested, used and stored. In many small mammals, some of these thermogenic responses occur in' tissue known as brown fat which – unlike the normal white fat that is a fairly inert storage tissue – can have a very high metabolic activity. It seems that this activity can be turned on and off as needed, and that animals can use brown fat either to produce extra heat when they are cold or to burn off extra energy when they are over-eating. The fact that brown fat evokes both these responses leads to an interesting interaction. If a rat is repeatedly chilled, it adapts by revving up brown fat activity to produce additional heat and, simultaneously, it becomes very resistant to weight gain. Conversely, if a rat is overfed on a cafeteria diet, it will not easily give way to cold.

The presence of energy use regulation in humans is less definite. In modern societies the simple compensations hardly operate. It remains true that, if you are overweight, you need more energy to run basal metabolism and get around, but much of the extra effort can be avoided by the studious use of chairs, cars and elevators. Similarly, finding plenty to eat is no longer matched by an increased energy cost, for loading more into the supermarket trolley has few metabolic sequelae. Some alterations in metabolic efficiency do occur in humans, although they are not as marked as in animals, probably because we have less brown fat.

Changes in energy efficiency are a familiar experience to many. If you try to lose weight through dieting, the body compensates by lowering your resting metabolism. For a couple of weeks you tend to get thinner quite quickly but after that the diet becomes ineffective. The reason for this is the suppression of both basal metabolism and your thermogenic responsiveness. It makes evolutionary sense. Your body detects that you are eating less than you are using but fails to understand that this is exactly what you had in mind. It interprets the situation as a time of relative famine and does all it can to help you get through it. Despite your wish to go on burning extra calories, it acts to block your plan by cutting down your needs. Soon you are so efficient that your diet becomes useless. It is a response designed to meet times of hardship

Although a less obvious experience, the opposite can occur if you consume more than you need. Basal metabolic rates and thermogenic responses then increase by a few per cent, making a big difference in an average sedentary person's overall daily expenditure, equating to the burning up of many calories. This change should dispel the myth that the overweight are cursed with a slow metabolism. In reality, they have the opposite, with considerably higher demands for energy than normal. However much it is denied, the overweight have to eat more than others simply to maintain their size.

Of course, everybody knows some fortunate people who can eat and eat without ever getting fat, and formal studies confirm their existence. Different individuals of the same age, sex, and activity levels may have calorie intakes that vary by as much as 40 per cent and yet are of similar weight and fatness. It is not clear why this should be so, but it could be due to the thin individuals having more brown fat than the rest of us, although the importance of brown fat in human weight control is often debated.

Some experts claim that the amount of human brown fat is so minimal that it cannot possibly influence our body weight regulation; others point out that you need to burn only a few more calories each hour, day in day out, to keep trim in the face of a considerable food excess. This seems logical to me, and leads to an interesting point. If brown fat is important in man, the overweight might benefit from being cooler in the same way as is seen in small mammals. Experiments have shown that when room temperatures are maintained at about 4°C lower than usual, subjects increase their energy requirements by a few per cent yet do not actually feel cold. However, the observation may have been due to the cool conditions making them move around more, rather than any change in brown fat activity. Nevertheless, it suggests that wearing less clothing and turning down the heating can help if you want to be closer to the lucky eat-all skinny types.

*

Although the level of the heating thermostat in our homes may slightly influence our daily energy expenditure, far greater changes are effected by our activity levels. Animals vary how much they do depending upon food supply, becoming less active when food is scarce and roaming more widely when there is a glut. Although effective in the wild, it is a feedback control that fails when they are domesticated. Just as the cafeteria diet can lead to rotund rats, spherical dogs can be created by owners who provide plenty of food but unnaturally few opportunities for exercise. It raises an important question. Which of these two processes – freely available dietary intake or low exercise levels – predominate when it comes to spherical humans?

After my long explanation of our failed intake controls, one might expect that people eat more now than they did in the past. This is not the case. Over the last twenty years, the average adult in Britain has decreased his or her daily intake by about 700 calories, and the figure would probably be much higher if we could make comparisons with the last century. Modern diets may have confused the controls set up to match our intake to needs, but those needs have declined so strikingly that, to my mind, today's large number of overweight men, women and children can be explained only by reduced activity levels. It has to be said that not all experts agree with this, and many claim that exercise has little influence on body weight.

It is true that even strenuous efforts can be maintained by surprisingly little extra food. A single small chocolate bar can fuel a run of many miles, and one would have to put in many hours a week at running or other sports to make much difference if the effects of exercise were limited to burning the little extra food required to undertake it. But this is far too simplistic a view. Following strenuous work, metabolism stays elevated for several hours, allowing considerably more calories to be burned off, and beyond that there are other effects on metabolism which lead to even greater benefit.

One of these is an increase in resting metabolic rate, which in most people normally accounts for around three quarters of their energy expenditure, and so a small percentage rise can be very useful for weight control. Performing regular activity stimulates muscles to become larger and stronger and, unlike fat, this increase in tissue will use energy even when doing nothing. After a few weeks of exercise your body is more metabolically active, burning more fuel throughout the day and night. The suppression of metabolic rate that usually accompanies dieting is also largely abolished, and your body is not so prone to enter the self-preservation famine mode even if you do go on restricting your food intake.

The influence of exercise is not confined to the output side of the equation, and this too can be useful if you wish to lose weight. When you first take up regular physical activity it will often suppress appetite, contributing to weight loss early in an exercise programme. I have frequently experienced this myself. Being of the type with a propensity to grow fat, I have found that exercising at meal times is a good way to regain control of my weight after a period when I have let things lapse. If I go for a run at lunchtime, my appetite disappears until well into the afternoon, by which time it does not take too much willpower to abstain from eating until the evening meal. If I don't run, I am ravenous and have to eat lunch. Unfortunately, this pattern does not last for long. Within a month of restarting exercise, I come back from the run very hungry, and this seems to be a common experience. It probably reflects increasing fitness. When you are unfit, your whole body's efforts are concentrated on keeping you going, and blood flow to the gut is minimised during your exertions. This turns appetite off. Become a little fitter and the gut receives a better share of the blood while you work. Then, instead of returning faintly nauseated, you come back ready to eat. Overall, people who regularly exercise eat more than those who do not.

Losing weight through exertion may seem disappointingly slow. If you go on a rigid diet, it is possible to shed pounds quite rapidly at first. If your body's needs are around 2,500 calories a day and you restrict yourself to 1,500, the deficit adds up to 7,000 calories a week, which is equivalent to more than a kilogramme (or two pounds) of weight loss, since both fat and muscle will be reduced. When you try to slim down through exercise alone, however, a number of factors operate against you. It requires a couple of vigorous exercise sessions every week to use up an extra 1,000 calories, and so it can take up to a couple of months to match the sort of deficit generated in one week of dieting. Furthermore, during that time, working out twice weekly will lead to a gain in muscle weight and an increase of up to half a litre in circulating blood. The latter will weigh 500 grammes in itself and so, overall, you can often find yourself losing no weight at all. This does not mean that you won't lose fat. In the last few months my wife, who has never been overweight but who was not as slim as she was in her youth, decided that she was taking too little exercise. She started going to the gym two or three times a week and has on occasions come running with me. Greater fitness has definitely resulted. The bathroom scales have not changed one jot, a fact she finds frustrating, but at the same time she has regained much of her youthful figure and is met by questions as to how she managed to lose so much weight.

Are diets, then, of any use at all? The answer must be yes, if rapid weight loss is your aim, but they should be used in addition to and not in place of exercise. Of course magazines, newspapers and bookstores are full of claims for diets that work wonders. Grapefruit diets, 'F-plan' diets, diets stopping you from mixing protein and carbohydrate – can they help to control weight? Here too I must also answer yes, but not for the bizarre reasons claimed. There is nothing extraordinary about consuming either grapefruits or exclusively high fibre foods, and the idea that you will benefit from keeping protein foods separate from carbohydrates is a myth. Most of the benefits are simply due to the imposition of restrictions which lead to lower intakes through both limiting choice and increasing consciousness of every item that you eat. They offer no magic, but a diet plan gives you strength.

*

So how – with job, family and social life – can a reasonable amount of exercise be fitted in to a modern lifestyle? Clearly many of us have too little time to be active for much of the day. We certainly cannot reach the levels for which evolution designed us. The question is: how little exertion can we get away with and yet retain our health and waistlines? Do we need to do vigorous training, or can we simply incorporate lower levels of activity into everyday life?

Young adults who do not have a manual job use about a quarter of their daily energy sleeping, with most of the rest divided about equally between sitting or engaging in light activities such as walking, housework, or shopping. Even if fairly enthusiastic, they spend less than five per cent of the time indulging in any sport or vigorous exercise. At that rate, even a doubling of sports activities will make little difference to the overall energy use. The best strategy for using up more calories would therefore seem to be changing some of the sitting time into light work. Two hours of easy gardening twice a week, for instance, is much the same in energy terms as running a half marathon. Most people could easily accommodate a little more movement into life without disrupting any routines.

For some years I worked in a laboratory complex that investigated all aspects of human performance. Many of the employees who were conducting research programmes for the British Armed Forces ran or played sports in their spare time, occupying an hour or two each week. Our headquarters had just ground, middle and upper floors and was equipped with one small and very slow elevator, situated at the far end of the building from the main entrance. It was also at the wrong end for collecting mail from the central registry, which was situated up two flights of stairs above the main door. It was with constant amazement that I would watch my sporting colleagues' choice of route. Every day I would witness one or another enter the building in front of me and turn to their right to walk the fifty or sixty yards to the elevator where, presumably, they pressed the button, waited for some time, entered the lift and waited again for the doors to close before the thing slid slowly upwards (perhaps stopping at the middle floor for some of the even more spectacularly infirm staff members). Finally they would reach the top floor to emerge and walk back along the top corridor to the registry itself – all to avoid climbing two floors. Even stranger, they would repeat the process to get back down again.

The same happens everywhere. At present I work in a large hospital with seven floors, and to walk up from the ground to the top is a demanding task – although still one I would encourage. But it is not those journeys that worry me. Just as in the research laboratories, all day I will see a variety of health professionals waiting by the overused lifts to travel just one or two floors up or down. It is very sad. Walking for just twenty minutes each day should yield weight reductions of two to four kilogrammes (4 to 9 pounds) over three or four months, and many of my hospital colleagues could achieve that within their daily work – and while saving time. The same, I am sure, is also true of many office workers. Most people could add some exercise to their daily routine, getting to and from their work place either on foot or by bike, or walking (instead of driving) to the local shop for that small forgotten item. Others could walk on escalators or those moving belts at airports, two places where standing seems even more mindless than waiting for the lifts.

Yet while this incorporation of movement into everyday life is important, I am not convinced that alone it is really adequate. From the point of view of fat control, it will certainly help to burn up more food, but there will be little effect on muscle size and advantageous changes in metabolism will be negligible. Exercise in life is not only about weight regulation. A little extra walking a day is very worthwhile but it is not going to grant a much stronger heart, lower blood pressure, improved HDL to LDL ratios and less sticky platelets. Our ancestors would have done more than merely walk, and so we too need some vigorous exercise if we want *all* the benefits. This, inevitably, takes commitment, but if you make the effort, in addition to staying thin, you will also see a healthier old age.

TEN

★

Fit for Life

HELEN KLEIN, the great-grandmother team mate of my first Eco-Challenge, may be exceptional by any standards but she is none the less an example of what can be achieved if you remain active into later years. After the race, Helen appeared in dozens of magazines and her remarkable abilities were broadcast throughout the United States, with reports on our team's progress screened on three consecutive nights on television's *NBC Dateline*. Inspired by her achievement, thousands of people across America realised that it was not too late to do something about their fitness. It was an awakening witnessed by the sheer scale of Helen's postbag.

Back in Britain, I told her story repeatedly, and that prompted an idea. My father, an active man approaching 70, had recently retired from full-time work. What if he were to undertake the next year's race? Perhaps he would provide some inspiration for people in Britain, and even if not, he would get a lot out of it. Several times while watching Helen hang from that rock wall I had wondered how my father might fare in her position. I did not know the answer, but was sure he would love trying. In many ways the idea was ridiculous. Whereas Helen was a world-class ultra-distance athlete, my father was a weekend walker, and had done no serious training for decades. Yet, unable to dispel the idea, I began to take it seriously.

We had been warned that the next course – to be held in the Rockies of Western Canada – would be much harder than it had been in Utah. Instead of ten days crossing deserts and canyons, the equally long route would cross glaciated mountains and would be completed in eight. When I explained this tentatively to my father, he was not deterred. A decision was made there and then to give it a go. It left me with a quandary – I needed to find three other team mates.

There were many factors to consider in making the choice. Each had to be capable of completing the race, and ideally some would have special skills. In particular, we needed someone who could read white-water, for this year the rapids would be run without guides. As before, each team had to include at least one female member and – most important – everyone had to be easy going. In Utah I had witnessed several teams destroyed by friction as fatigue and tetchiness set in. Furthermore, we all had to accept that we were not out to win, for with my father on the team that would be impossible. Finishing the course had to be the challenge in itself.

Ranulph Fiennes was my first approach: it would be great to do something together again. He was also well known, which would help us to gain publicity and sponsorship, and I recalled that he had been a canoeing instructor for the army and so should know about white water. Next, I spoke to Rebecca Stephens. I did not know her personally but she had been the first English woman to climb Mount Everest, and after that she had gone on to ascend the highest peak of every other continent and to complete the 'Seven Summits'. She had since become a presenter on BBC television's popular science series, *Tomorrow's World,* and so she too came with a celebrity badge. Both Ran and Rebecca immediately agreed to join us. Now we were four, looking for a fifth. It was then that something struck me. I wrote down the ages of my team members so far – Rebecca in her 30s, me in my 40s, Ran in his 50s and my father in his 70s. If we could find someone in their 60s, we would have a real pattern, and consecutive decades would give an enormous boost to the promotion of fitness for all ages. What we needed now was a strong, determined 60-year-old. But who would fit the bill?

A few months earlier I made a train journey back to London from Leeds, dining with Chris Brasher, the chairman of the cold weather conference we had both attended. Chris was an Olympic gold medallist from many years back and had founded the London Marathon. I knew he remained heavily involved in athletic and outdoor activities and that he still participated in long fell races. He had to be in his 60s, and he not only confirmed my guess but agreed to join the project. Team 'Fit for Life' was born.

*

Just downstream from the town of Lillooet, the river has carved a huge straight valley through the Rockies, its floor covered in a patchwork of fields and homesteads. Above them, the towering valley sides are cloaked in dense coniferous forest, leading up to dramatic glacial peaks. Here the second Eco-Challenge was to begin – a twenty-five-mile stage along the river banks. As in Utah, we were to start with both riders and runners, but this time we had only two horses per team.

Waiting for our mounts in half darkness and early morning mist, I was nervous about the ride ahead. My team was as strong as it could be, although not quite as I had envisaged. My hope for a balanced representation of age had had to be foregone. Chris Brasher had not appreciated how much preparation would be needed, and when he realised exactly what this entailed, he decided that participation was unrealistic. When he finally dropped out, with just a few months to go, it was too late to find another 60-year-old who could train up in time. I turned to a doctor friend, David Smith, who like myself was in his 40s. It rather spoilt the pattern, but except for Rebecca, we all remained far older than the majority of other entrants. My father, of course, was the oldest man in the race, but he was not the oldest participant. With another team, Helen, now 73, had come back for more.

The obvious pair to ride the horses were my father and Rebecca, for Ran, David and I were used to running long distances. If we did the opening marathon on foot, the team as a whole would be in better shape for Stage 2. All the same, after Utah I appreciated that riding such a distance is not effortless. My father had never been on a horse in his life before he entered for this race. He had hurriedly taken a few riding lessons but I had no idea how he had got on.

Fearing another stampede at the start, I begged my father to rein back his horse across the first half-mile field. It was a pointless suggestion. Although the horses were better schooled than those in Utah, they were still frisky with the excitement, the dawn, the noise, and the numbers. As the countdown finished they were away and I could do nothing but watch with dismay as my father galloped across the meadow, looking pretty unstable on an unfamiliar western saddle. I was sure he would fall and be badly hurt, even if he was not trampled. But he did not come to grief. At the far end of the field, where the route went through a gate and up on to the top of one of the river levees, a bottle-neck brought everyone back to a trot or a walk. It gave us a chance to re-form, and we moved up the bank as an organised team.

The next hour or two went well. The horses maintained a steady trot, as long as the riders kept a tight grip on the reins, and we found an easy loping pace. At the halfway watering station our animals were passed without comment, though the vet paused to look at my father, a question written across his face. Was this man, already hobbling, fit for the tasks ahead?

I too was anxious. Although confident enough to set out in the saddle, my father had not become an accomplished horseman. It was painful to watch his bouncing progress on a high-trotting mount, and when he got off, he was obviously in great discomfort. He ruefully admitted that his backside was already badly beaten and blistered. Clearly he needed a break, but if he had one, we would slow down considerably. Shamelessly I asked if he could continue in the saddle, and although his face dropped, he did not complain. Slowly, stiffly, he remounted, grimacing as he returned to torture.

His steadfastness lasted for another hour, by which time we had reached the last quarter of the run. Then, suffering from cramps, he simply had to dismount. David took the horse while father, obviously tired, began walking and then, after loosening up, broke into a jog. Over the next few miles we maintained a steady pace, Rebecca on one horse and David, my father or I taking turns on the other. Ran stayed on foot, sometimes disappearing ahead in his enthusiasm. A lifetime of competitiveness made it hard for him to hold back.

At the end of this ride/run stage, our route had to cross the river, even though after heavy rains, it was in full flood. This made it far too dangerous for the horses to cross, but that did not mean we would be spared the challenge. We handed our steeds to some wranglers and dropped down to the water through a small wood, emerging from the trees close to the bank. Nothing would have prepared me for what lay before us. Instead of the peaceful river of a few miles back, the waters had left the constraints of the levees and had become a violent cauldron of grey-white power. We saw from the efforts of a team ahead – all young strong tri-athletes – that crossing it was to be a ferocious battle. With the water just a little above freezing, hypothermia was a great danger and my father would be particularly vulnerable. Unlike most older people, he had never gained fat with age.

*

It is accepted as inevitable by many that the older you become, the more fat you will carry, even if you do not put on any weight. I am not convinced. Although I accept that it is the norm, I do not believe it to be a natural consequence of ageing. It is more a reflection of declining activity.

The tendency to increase internal fat with age was first demonstrated in studies comparing measurements of the thickness of fat layers beneath people's skin, with estimates of their whole body fat made from underwater weighing assessments of body density. The results showed that although both skin and whole body fat tended to increase with age, the latter changes were more marked. Indeed, there was even an increase in whole body fat in subjects who showed no change in skin fat thickness at all. Some of the elderly subjects looked slim, but they had more internal fat than the young. It seemed that muscles and other organs developed higher fat contents with increasing years, even in those who did not put on much weight.

As a result of many such measurements, scientists were able to write equations to predict whole body fat from skin measures alone, with varying co-efficients to account for age. Equal measurements of skin thickness in a young as opposed to an older subject would therefore predict that the older had a higher total body fat. Such equations have been used widely in research, but when I first saw them, it struck me that the lower activity levels of older men and women might explain at least a part of this apparent age-related change, and these suspicions were supported by measurements of skin and body fat made on Ran Fiennes and me before our Antarctic crossing.

At the time of our crossing, Ran was in his late 40s and so, according to conventional dogma, should have been internalising fat at a steady rate. When I measured his skinfolds, they corresponded to the expectations one had when looking at the man – small pinches of a lean, fit individual. But the skinfold to body fat conversion equations suggested that, at his age, this limited skin fat thickness still reflected a total body fat of more than 24 per cent. I doubted this, and we had underwater weighing measurements which were very different. In reality, Ran was still a lean 16 per cent fat, with a relationship between his skin fat thickness and internal body fat closer to that expected in a 20-year-old. To my way of thinking, the reason was self-evident. He had refused to become sedentary. Instead of depositing fat in his muscles and other organs, he was still using it for the purpose nature intended – refuelling his regular exertions.

These results prompted me to examine others who might have delayed their 'age-related' changes. I looked at skin and total body fat measurements in a number of active men in their 40s and compared the results with men of the same age, weight and height who were essentially sedentary. The results were as I anticipated. The fitter men had similar or less fat under their skin than the inactive, but much less total body fat. Clearly exercise can alter an accepted part of ageing.

*

We had been warned that if there were any weak swimmers in the team, it would be advisable to have a safety line for fording the river, and I was carrying a full length of climbing rope for the purpose. My intention had been to rig it right over the river but now I realised this was impossible. Even a full length climbing rope was far too short, and it was clear from the team before us that the whole width was to be hard swimming. So fast was the water running that as competitors jumped in they were picked up like flotsam. Nobody made it to the far bank within three hundred yards, let alone within a fifty yard rope span. There was, however, one hope. Trapped in midstream, about a hundred yards down river, was a large log. It would be impossible to reach it from this bank, but it looked as if one could get to it from upstream on the far side. I decided that if I swam the river first with my kit and the rope, I could work my way back up the far bank and then swim and drift down to the log. From there I should be able to throw a line to my father as he went past.

My worst fears of the water were nought compared to the reality. As I leapt from the bank I was engulfed in the raging torrent, and it was so cold that it crushed all breath away. It took enormous will to concentrate and strike out for the far bank, a difficult task when washed sideways in a liquid roller coaster. Progress was also hampered by my pulling the kit and rope in a large waterproof sailing bag which dragged terribly. Eventually, however, my feet touched bottom near the far side. I was then reminded why we had been advised to keep on swimming until in real shallows for, with the water running thigh deep, I was simply knocked back down. By the time I emerged on the far beach, I was an exhausted cold man, and I am neither thin nor weak when it comes to swimming.

After resting for a couple of minutes, I took out the rope and set off up the broad, shingled bank until I was much higher upstream than the log. Gritting my teeth once more, I then re-entered the freezing torrent and allowed myself to be carried down, making my way towards it. The plan worked and I got there, but each time I tried to climb on to the dead tree, the power of the river's flow and the slipperiness of the surface defeated me. With my legs and chest screaming from the cold, I simply had to get out of the water soon. So I released the log and swam back to the beach. Shattered and shaking, I put down the now sodden rope and waved to my father to come on. He would have seen my failed attempts and must realise that I could do nothing. Yes . . . he and David were heading down for the water.

They jumped simultaneously. I saw David strike out strongly for my bank while my father appeared to do nothing. There was little sign of movement and certainly no progress – just a bobbing head racing down the river a few yards out from the far bank. With both his own and my father's kit, David was already too far away to help. Another few seconds and my father would be passing where I stood but still the whole river's breadth away. I had seen around the next corner that, less than half a mile on, the river narrowed and steepened into a boiling boulder run. I had to do something. A little downstream, there was a small spit that stuck out into the flow. I sprinted down and out along it, diving from the end into the swirling rapid that it generated. The rapid pushed me out towards midstream, where I pitched into hard front-crawl, trying to head across the flow. For thirty seconds I did not look up. I knew it was a race to make contact quickly and that it would take at least that long to reach my father's stretch of water. If I were too slow, we would both be in trouble. The water was too cold to sustain activity. If we were still in it when we reached the rapid, we would be unable to defend ourselves at all.

Powering on as best I could, I literally counted the seconds, face screaming from swimming with my head down. When I reached thirty, I changed to breast stroke and tried to look around. I was perhaps two thirds of the way across the deep segment of water, in big waves and running down the inside bend. With the water in such turmoil, it was difficult to see any distance. For a few seconds I thought that I must have completely mis-timed things and that my father had already passed. Then I caught sight of him just a few yards away and roughly level. He was staring straight at me, but he made little sign of recognition or movement. He looked for all the world like a man day-dreaming in the bath rather than one about to go down the plug.

I reached him in just a few strokes, grabbing at his collar and immediately setting off back towards the far beach. The nearer bank was too steep and inaccessible. It had been years since I did my lifesaving course at school, but I remembered how to pull a man to safety while maintaining his airway. My father kept getting a faceful of water, but it made him cough, so he was obviously all right. Progress was desperately slow and became slower still as fatigue and chill poisoned my leg muscles. It was so cold around my chest that I could scarcely breathe. Then, just as I thought I could go on no more, my foot touched bottom. Simultaneously, I also noticed another figure in the river below us. Roped to the shore and wearing a black rubber suit, the man was clearly a safety backstop. It was welcome reassurance, but I did not want his help. Rescue would mean disqualification.

Stupidly, I tried to stand as I had done before but was again reminded of the futility. I went under and in my tiredness lost grip on my father.

He began to drift away but, coming to my senses and revitalised by the shallower water, I grabbed him again and made another final effort. It brought us into shallows where I could wade, and my father also made an effort to stand. We could not help the river from pushing us on downstream, but we were now out of trouble. In the end, we were helped over the final few yards by the man in black, but I had done my bit and had no qualms about his assistance.

Of course I should have realised that help would be at hand. The race organisers knew there might be problems with such violent water and had strung a cable across the water above the rapids. In the heat (or perhaps cold) of the moment, the presence of such a safety net had not occurred to me when I saw my father heading for serious trouble and needing help. In the event, I need not have bothered. So many competitors had to be rescued that the rules about disqualification were abandoned.

*

At the time we decided to enter the race, my father had about eight months left for training. He started immediately and, a few weeks later, when I was over at his home for the weekend, I suggested a run together. We went out early one chilly and wet Sunday morning, with the trees bare and paths sodden through field and wood. The mud made the going hard, but I was still dismayed when, after only a mile or so, my father had to stop on a slight incline and walk. He was carrying nothing, we had not gone far, yet he was out of breath. Had I made a dreadful mistake?

It was too late to change my mind and steadily the weeks and months went by. I tried to organise dates for a few weekends when we could all go to South Wales and have access to canoes, rock walls, horses, mountain bikes and open country. Practice for glaciers and white-water rafting was not going to be possible, but I felt we could rely on my own skills for the former and Ran's for the latter. Unfortunately, when I spoke to Ran about this, he denied all knowledge of rapids and suggested that, as I had run the Colorado the year before, raft captain should also be my job. In any case, he had no time to spare for the weekend sessions and in the end made none of them. While Rebecca, David, my father and I spent at least some time hanging from cliffs or paddling up rivers, Ran's refresher course in abseiling and climbing was held on a rope slung from the chimney of my house. It was hardly ideal training for big rock walls, but it had to do.

If concern remained over aspects of our specialised preparation, I was encouraged by my father's progress. He committed himself in earnest, and it was great to see him paddling a canoe up the river Wye all morning, battling on ropes up a huge cliff all afternoon, and then tramping over the hills as the sun went down. The transformation in his fitness was also astonishing.

*

After staggering from the Lillooet, it was some time before my father and I were in a fit state to continue. Besides the physical exhaustion and the cold, we were both quite shocked by what had happened. Indeed, the events in the river were to replay themselves in my mind with vivid intensity several times over the next few days. It was very disturbing. Had I been grossly irresponsible?

From the river bank to the first stage checkpoint was only a few hundred yards. There our support crew – Morag, who had been the radio operator for Ran and me in Antarctica, and her daughter Moyra – gave us a late lunch while we studied the next part of the course. The huge climbs over glaciated peaks and passes, separated by lower segments through wooded slopes and valleys, would obviously be taxing but the terrain did not look too bad. We reckoned we could complete the stage in a couple of days and nights.

It was still hot when we left the checkpoint at around 4 p.m. and made our way up and into a side valley. The lower slopes fell precipitously into a narrow river gorge, and so we had to climb more than three thousand feet in order to contour in above the cliffs. There was a logging track to help us make the climb, which would have been easy had we been fresh but, after a day which had seen us run a marathon and then battle with the river, the hairpinned route was hard work, and we baked in the heat. Darkness was falling before we had finished the climb and were ready to enter the forests.

Here, orange flags were meant to mark the way, but under the dark canopy we soon lost sight of them. Traversing across the steep hillside, there were the prints of boots and trainers all over the place under the trees where teams had cast up and down in their attempts to pick up the trail. We decided to maintain height by sticking to our altimeter, but the slope was often cliffed and we were forced to climb or descend to find ways onward. At around midnight, with the slope so steep that we had to hold roots and branches in order to make headway, we decided to stop and rest. Progress would be safer and faster in daylight, and our water was in short supply. Below us, we could hear the river raging. It sounded quite close, and so we dropped down to find a place to stop. We had not gone far when we reached a broad ledge of rock and mossy earth large enough to accommodate us. The spot was somewhat spoiled by a dead tree trunk lying across the middle. Beyond the ledge was steeper rock – perhaps a small cliff down to the river itself.

We sat down on the tree trunk and nibbled some bits of food without cooking as our rations needed rehydration. We then made more room by heaving the tree from the ledge. It slipped easily into the darkness but, to our amazement, several seconds passed before we heard its distant crash. Clearly the cliff was higher than we thought, and we all moved a little further from the edge. As we lay down to sleep, Rebecca, who was close by Ran, took out her ice axe and planted it firmly between them. She saw him eye its vicious blade nervously.

'Don't worry, I do trust you,' she said with a smile. 'It's for the bears, not you.'

*

After the age of twenty-five or so, it is accepted that your aerobic capacity will decline by about one per cent per year. It means that by the time you are fifty, you will be very much less fit than when you were in your prime. Although this is the accepted norm, many studies have shown it need not be so. The rate of decline in the aerobic capacity of older endurance athletes who continue to compete is only half as fast as normal, and their decrease does not begin until well into their thirties. Indeed, a couple of studies on elite veteran athletes, who have gone on training into middle age at the same sort of intensity as when they were younger, show no decline at all until they approach their sixties. As they aged, they became no more breathless as they ran, and there is no reason at all why we too cannot limit our declines in this way.

Even if you have not maintained activity throughout your life, but come to it when older, the benefits from training can be quite extraordinary. My father's improvements over the eight months leading up to the Eco-Challenge provide a good example of the benefits to endurance performance, but even more marked improvements can accrue when it comes to strength. Even at the age of eighty, people can benefit as much as young adults if they take on a strength-building programme. This is not to say that they become equally strong but that they improve by a similar proportion. A twenty-year-old who has done little training can double his or her strength in just a few weeks. So can the elderly, and in their case it makes a greater difference. Instead of a change that simply improves sporting prowess, increased strength in older men and women can work miracles. Much of the unsteadiness of age is simply down to weaker musculature, and training can literally put people back on their feet. Overall, it can take the person who needs help when getting out of the bath to a life of more independence and dignity, the person who can only get around the house down to the shops, and the person who can make it to the shops up into the hills. There is no doubt, exercise can be a fountain of youth.

*

My watch alarm woke me at 5 a.m. – less than four hours after I had dozed off. It was still dark but the moon illuminated the scene. I lay there for a minute or two, savouring the moment. It was great to be out in the mountains, surrounded by awesome beauty, and good to be there with my father. We had to move with daylight, and so I woke the others and made ready to depart. While we dressed, my father asked me to have a look at the sores he had acquired from the riding on the day before. I thought he was being fussy until I saw them. Then I was horrified. His backside had been rubbed by the saddle until areas the size of saucers were left bloody, raw and devoid of skin. It was obvious that they would not heal quickly and would give him hell, but he would not hear of stopping. I covered them as best I could with adherent pads, but I knew the wounds would weep so much that the dressings would soon become useless.

By the time we were ready, there was enough light to see down the drop over which we had sent the tree and, for the first time, our precarious position was brought home to us. We were camped on an overhanging ledge, high up on a gigantic cliff, hundreds of feet above the river. What was more, much of the ledge consisted not of rock or solid ground but had been created by some rotting tree trunks falling and jamming across rock pinnacles. Indeed, you could look right through gaps in the floor of this giant nest, and from the hollow which Ran had chosen for comfort, you could drop pebbles straight to the water far below. We abandoned our plans for losing height and finding water, leaving our spectacular eyrie by the way we had come.

A few hundred feet higher, we turned the top of the cliff and moved along the slope, much as we had done in darkness the night before. After an hour or more, the whole valley widened and the gradient became easier. The vegetation, however, became increasingly difficult to negotiate. I had always thought of coniferous forests as bare straight trunks rising to a canopy with little undergrowth. The early part of the woods had been just that, with trees of all ages – some young, thin and tender; others ancient and of enormous girth and height. As well as those that lived, there were many that had died and dead branches and trunks were strewn on the forest floor. Some were so rotten that they shredded beneath a booted foot or were nothing but a line of decomposing dust. The whole of life and growth, death and decay was written clear on the woodland floor and the place had an air of primaeval mystery.

As the slope eased, the nature of the forest changed with the addition of vibrant lush green undergrowth. Two plants were particularly common, both designed to make life difficult. One had numerous thin branches that grew like a horizontal lattice to bar progress. Our feet constantly caught upon them, whereupon their bark stripped instantly, turning them from straightforward obstacles into slippery wands that lay across the slope. They were aptly named slide alder, and encounters often brought a crashing fall and a short roll downhill to meet the Devils Claw – a plant with barbs so sharp that they went straight through our clothing. The Devils Claw grew avidly beside the slide alder and was always the plant that one grabbed to try to stop a fall. We took wide detours to try to avoid these evil plantations, often walking along giant fallen trunks well clear of the ground. Even so, before long we all had cuts, scratches and hands filled with thorns, and to add to the torment, vicious hornets stung our faces. It did make me wonder why we were there.

We struggled through the vegetation all morning, stopping once as soon as we could drop down to the river to have some rehydrated breakfast, supplemented by welcome wild berries that tasted like redcurrants. Then, at around midday, and with increasing altitude, the vegetation began to thin out and we came across the orange flagged trail once more. It made the going easier, but within a few hundred yards we reached a point at which the valley branched. The orange markers went straight on, while the map suggested that we should turn right. Reluctant to leave a proper trail, we decided to follow the marks, believing that they must swing round some yet invisible obstacles. This meant fording two large rivers, wading thigh deep across the fast flowing torrents. It was a bitter experience, made even more humiliating when we discovered that the flagged trail finished just a hundred yards further on at a hunter's cabin. The path was nothing to do with the Eco-Challenge course, and miserably we had to accept our mistake and recross the icy waters.

We spent the next few hours slowly gaining height, either fighting with plants or climbing up screes, finally to exit all vegetation in a high snow-filled valley. It had taken the best part of twenty-four hours to get there, despite being little more than twelve miles on the map, a rate of progress that did not bode well for the remainder of the stage.

At the next checkpoint, we learned that more than two thirds of the field were still behind us. The news gave us considerable encouragement for clearly others were also having problems. Later we heard that some teams had become so lost in the forests it took them a week to make that point.

Here David confided that he was feeling dreadful. Perched on a rock beneath the checkpoint, he was clearly in a bad way – shivering, despite strong late afternoon sunshine, and looking extremely pale. A lack of food and perhaps hypoglycaemia seemed the likely cause, and so we halted for long enough to make a brew and cook a meal. The diagnosis proved correct and David recovered rapidly.

The route continued upwards to cross a col before dropping steeply down a glacier. Our second night fell as we descended, and at around 9 p.m. we left the ice and skirted a glacial lake. There I called a halt. It was useless to become exhausted so early in the race, and we were about to re-enter the forests. Tackling the plants in darkness would be a nightmare. Instead, we lay down on the moraine and went to sleep.

The following day, the third of the race, began with another six hours of forest-bashing before we finally approached the bottom of a valley, where we needed to cross the river. It was clear from the map that this was likely to equal the size of the Lillooet, and my heart sank at the thought. If it was to be anything like the Lillooet crossing, I was not prepared to try it. One close call with my father might be considered bad luck, another would definitely seem careless. Fortunately, I need not have worried. The river had been rigged with a Tyrolean rope traverse, and we all had the excitement of working our way across, dangling inverted a few feet above roaring rapids.

Beyond the crossing was another enormous climb, four thousand feet on a single unbroken slope. In the hot afternoon sunshine, it was very hard work, and soon Ran, David, my father and I were sweating wrecks. Rebecca, on the other hand, remained cool and composed. Throughout the race she was to demonstrate extraordinary physical resilience, coupled with humour that was never bowed by fatigue. She had an amazing ability to look more like a model on a fashion shoot than a competitor in one of the world's toughest endurance races. Beyond doubt, she was a tremendous asset to our team.

The climb took us from forest to pasture, to scree, to ice, with the vista behind becoming ever larger. Finally we arrived on a high alpine summit, surrounded by glorious views. We stopped and savoured the mixture of effort, achievement and beauty – what being in mountains is all about. While sitting on the peak, we reviewed the rest of the hiking section. It made us troubled. We had greatly underestimated the difficulty of the forests, and now the two nights and days that we had thought would take us to the next checkpoint had passed. There was still a long way to go with barely any food.

During the evening we descended across large snowfields, but we were still high as darkness approached. Turning the end of an ice-bound ridge, we came to a steep drop. It looked a dicey venture in the dark, especially after fourteen hours of travelling. I glanced at my father and saw that he was limping. I did not know if this was due merely to fatigue or whether he had pain from his hip – a considerable problem in the past. Either way, to continue seemed unwise and I called another halt. After we had settled, I asked him about it. His reply reassured me. Although tired, his hip was fine. Indeed, it was giving him less trouble than it had done for years.

*

A common criticism of too much exercise, especially as you get older, is that it leads to damage from long term wear and tear. This is largely a fallacy. In itself, exercise strengthens bones, joints and cartilage as much as muscles. Naturally, if you are untrained and overdo things, it is easy to wreak terrible damage to joint surfaces that are not hardened and ready for it. The time taken for bones and joints to adapt to exertion is longer than that taken by the muscles that move them. Over enthusiasm can therefore do harm, and I have been guilty of this many times. Even when not in training, I know that I can run twenty miles or more on determination alone, but while determination can drive muscles, it cannot take care of joints and tendons. Suddenly returning to exercise with another expedition to prepare for, I have paid the price, and there is nothing more frustrating than being forced to rest a sore ankle or knee when wishing to get fit.

Although injuries are unavoidable in sports, some later infirmities are unfairly attributed to them. There are many ex-athletes with painful ankles, knees and hips, for the same reason as there are many in the sedentary population with the same complaints. Training does not remove the possibility that someone might suffer from arthritis. The trouble is that the former activities of athletes invariably take the blame, and so the myth that exercise causes joint damage is perpetuated. The reality is probably the opposite. Exercise makes it less likely that you will end with problems, and individuals may well suffer worse invalidity if they never participate in strengthening activities.

Broken hips, wrists and ankles are common in the elderly because their bones become so very thin. Although continued exercise cannot protect the skeleton completely, it can largely eliminate the chances of fractures. People tend to think of bone as being a static, inanimate substance but the reality is far from this. Bones consist of a living matrix of connective tissues with a blood supply and nerves, and contain millions of living cells that produce the matrix and the hard calcified mineral deposits which are superimposed upon it. These cells constantly work to fashion the bones in order to meet the demands put upon them. Some build strength to make the bones more dense when required, while others do precisely the opposite, weakening the structure if demands are few so that the body's resources are not wasted and the skeleton remains as light as possible. The system worked well for millions of years but, unfortunately, it no longer does so. As with so much of our physiology, it was set up to expect daily activity and so the bone building system needs almost continuous stimulation in order to do its job properly. In many modern lifestyles, this simply does not happen.

The man or woman who drives to work, sits all day, drives home, and then watches television puts little stress on their skeleton and, inexorably, bones that were strong at the end of teenage growth begin to waste. It takes decades to become serious, and so it is in later life that disaster strikes. A trip, a twist and a fracture results. We all know people who have done it, but few realise that most cases could have been avoided. The problem is particularly bad for women. The balance of bone deposition and destruction is also influenced by hormones. Where males continue to get some stimulation to bone growth through middle and older age, females lose that stimulus with the menopause. It is therefore vital that all women approach the change of life with as dense a bone structure as possible. Although a healthy diet can help, there is only one way to ensure this. Women, even more than men, must continue exercise beyond their schooling years.

*

In many ways the fourth day of the race was similar to the third. We spent much of it at altitude, crossing a series of glaciers. We were roped at times, but there were no great difficulties, and it was not until early evening that the routine plodding was broken by our arrival at a high alpine ridge where another checkpoint nestled in front of two enormous upright stones. We signed in and passed on, having been warned that the descent beyond should be taken with extreme caution. They were not joking. Passing through the giant gateway – a natural equivalent to a cromlech from Stonehenge – we looked down a slope of broken chaos. A huge scree-field, bereft of snow and littered with gigantic rocks, stretched away below us – a mountainside laid waste, with huge angular boulders piled upon one another. It was a nightmare to descend for the rocks were loose and very sharp to hold. Several times they would shift beneath our feet, threatening us with a cascade of mammoth blades of granite that would crush a limb with ease. We were grateful when we reached the bottom unharmed.

From the foot of the scree we descended once more on a glacier. The further down we went, and the warmer it became, the more the surface was covered in meltwater streams which ran along before tumbling into deep cracks in the ice. Soon the whole surface of the glacier was more like limestone caste, and our travel was broken by the need to jump over the ever enlarging cracks, or edge gingerly round the more daunting caverns. The worst difficulty, however, was saved for the last. Without thinking, we cut to the right hand side of the glacier when not much further down the valley we had to climb out via the left. Bursting from beneath the snout was a large river – grey, cold and filled with glacial fragments. Once more, we were on the wrong side, and the day ended with another deeply chilling fight with iced waters.

After a third uncomfortable night, we passed down through a mixture of empty floored conifer forest and sunny grass-filled glades where redcurrant berries grew prolifically. Though we picked them freely to stave off our hunger, we were always wary, for the berries are the staple diet of the brown bear. Fears of bear attack had been with us from the start, despite the fact that we carried a noxious spray and were told that a noisy passage would keep the bears away. To this end, we wore around our ankles little golden bells that jangled as we moved along, but they hardly gave us confidence. For that matter, neither did the advice that we should lie still and play dead if attacked by one of these ten-foot teddies. It was in all the books, but we had no confidence that the bears had read them. Furthermore, you were meant to behave differently if your close encounter was with a 'grizzly'. Quite how, in the heat of an attack, you were supposed to tell the difference between a brown and grizzly bear was not clear. The best way was apparently through examining their droppings. Brown bear shit would contain obvious berries, while grizzly shit would contain bear bells!

Lower down, the forest thickened up again, and we had more fun and games with the vicious plants and hornets. Still, we made good progress until, late in the morning, we met a large and fast river coming in from a side valley. A short distance upstream, the river was spanned by a fallen tree, a log running high across the waters, starting broad and strong on our side but becoming disturbingly narrow before it reached the further bank. There was no other way to cross. One by one we climbed on to the unstable bridge.

When it came to my turn, I was surprised to find how difficult it was to walk along the log. I looked down and found the moving waters almost hypnotic. It took little imagination to know one's fate if one slipped. As the trunk became thinner, there was a marked increase in vibration, and the whole tree bounced up and down. Reaching the far side, I was fearful for my father and shouted back, asking if he wanted a rope. He seemed confident that he could do it without, and very slowly, with care but no hesitation, he started on his way, moving step by step with an expression of infinite concentration and self control, hiding the fears that must have welled up inside him. It was a savage test of nerve. Then, as the last few feet approached, his features began to relax and transformed into a smile of triumph and satisfaction, while we others burst into spontaneous applause.

How we all had changed. A few days back, none of us would have dreamed of making that crossing unroped, but now we all had confidence in our strength and in each other. We had become close, and we urged each other on where necessary, or held ourselves back to stay in a group. We supported ourselves with humour and camaraderie that forged strong bonds. To all intents and purposes, we had become the modern equivalent of a small roaming band – not hunting down some quarry but deriving pleasure from work as a team over days of physical hardship. Perhaps we were dipping into ancient race memories.

*

As we approached the second staging point in the late afternoon we were surrounded by photographers, reporters and TV crews – all hurling questions, mainly at my father. 'Say, how're you doing?' 'Are you hungry, or just tired?' 'D'ya reckon you can keep on truckin?' Dismayed by his rather understated answers – 'Fine', 'No' and 'Yes' – they turned our way, displaying the peculiar American obsession with team psychology and jargon.

'How does the interactive dynamic within a team containing inter-generational family members affect your role?' was one gem I heard aimed at Ran.

'It's great,' he responded smoothly. 'Whenever Vic tries to drown himself in a river, I know that I needn't do anything about it.' The interviewer was suitably perplexed.

At the checkpoint we contemplated the next event over a hasty meal – canoeing two long segments of ribbon lake, broken by portage around a dam. The whole thing was likely to take twenty hours, but there was no question of sleeping. We made our preparations as quickly as possible, but darkness had long since fallen when we made our way down to the lake shore.

Casting off and paddling into the night was tremendous. There was no wind, and before long the moon rose. I travelled with Ran and my father in one canoe while David and Rebecca took the other. It was very peaceful, and the regular dipping of the paddles was the only sound to be heard as we went gliding along. Progress was swift, and within a few hours we came to the dam where we had to put in and travel overland. It took some time to get the two heavy canoes – almost 90 kilos (200 pounds) each, including our kit – up the steep bank to where a narrow road led us a mile or so to the point at which the river could be rejoined. Dragging the boats was forbidden, so we had to carry them one at a time along the dark and deserted road, stopping frequently to rest our screaming muscles. They were not only heavy but very awkward.

Beyond the dam we faced an interesting choice. We could either put the boats into the water close by, and then run some rapids, or we could carry them an extra few hundred yards and miss the worst of the white water. With no misconceptions about our abilities, we opted for the latter and carried the first boat down to the second put-in. There we left it by the side of the road, not far from a small village associated with the dam and power station. My father opted to stay and watch over it when the rest of us set off to fetch the other. While we were away, he decided to create a comfortable berth from the kit in the canoe and settled down to sleep. When we arrived back at dawn, we met a woman walking her dog and a TV crew out filming the Eco-Challenge. Both were delighted by the cameo. Even in Canada, it is rare to find a grey-haired 70-year-old dossing in a canoe by the roadside while a woman walks past with her dog.

We were about to embark once more when some bright bags floating down river caught our eyes. Not far behind them came first one and then the second canoe of another team, both upturned with the crew trying to cling on to the hulls as they swept by. We waved gaily to them but, perhaps not surprisingly, elicited no response. It certainly confirmed that it had been a good choice to avoid the upper rapids. Mind you, the river was not exactly placid where we cast off. All eyes were watching for concealed rocks when suddenly Ran, my father and I realised that we were heading for a sandbar in midstream, just covered by water. Desperately, we tried to take evasive action, turning across the flow and paddling like demons, but it was not to be avoided. Here Ran's denial of canoeing experience was shown to be false. As the boat approached the shallows, sideways on and soon to run aground and turn over, he leapt out, turned the boat head on, pushed it over the bar and jumped straight back in again. It took him no longer than two or three seconds, so David and Rebecca assured me. At the front of the boat, I had seen nothing of what had been going on behind and so I was mystified by our abrupt change in direction and escape from capsizing.

Beyond the rapids the river entered a second long ribbon lake, down which we would paddle for most of the rest of the day. At about lunchtime we stopped briefly, pulling into the shore to show our race card at another checkpoint. We debated sleeping awhile, with Rebecca keen to rest for an hour or two. We had now gone for well over thirty hours without sleep, and before that for several days with only brief uncomfortable stops. Besides, other than my father who had snatched some shut-eye in the boat, we had begun to hallucinate.

Paddling along in a half reverie, we had all been seeing strange things in water, hills or sky – not true hallucinations in the sense that one believed in them but still very real. I had seen huge words written across forest hillsides and some gigantic animal heads carved in the cliffs. It struck me as odd that I couldn't easily make them disappear, even though I knew that they could not be there. Yet they were not unwelcome. Although mildly disturbing, they made the time go by.

Hallucinating or not, I was still concerned about our position in the race. We remained in the top third of the field, but we would be slow on the mountain bike section to come compared to many of the teams behind us. The end of that section would be a critical cut off with many teams prevented from continuing if it was felt that they could not complete the course before the eight-day deadline. I felt sure that we couldn't afford to stop at all, even though it meant another wakeful night to come. I insisted we should continue.

At around six that evening, having paddled for eighteen hours non-stop, we reached the end of the lake and the next staging point. The setting sun cast a glorious light across the water as we cooked and ate. With tired limbs and backs, each of us thought about remaining there – resting, relaxing, and enjoying the wonderful surroundings instead of meeting them head on. But the race officials confirmed my earlier fears. We had to get through the biking section by sundown the next day or our race would be over. Wearily, we picked up our bikes and packed to leave.

The heavy rucksacks on our backs – containing food, fuel, camping gear, safety ropes, and extra clothing – made cycling extremely difficult. Things grew worse when we met the start of the hill, a hairpin logging track that would take us up about three thousand feet and over a mountain pass. It soon became evident that pedalling uphill was not an option with the weights we carried and so, dismounting, we reluctantly began the long push just as night fell.

It took more than four hours to make that climb, and winding back and forth through essentially unchanging forest might have been boring were it not for the hallucinations that altogether eclipsed our earlier experiences. Exhausted from physical effort and lack of sleep, we found our minds entering a wild, free-running state of continuous delusion. All around us as we walked were men, animals, and cities, created from the play of moonlight on trees and rocks, but to our thinking, as solid as concrete. For some reason, I saw many giant Disney characters, or at least things quite like them, and they were not just transient. Some I saw from hundreds of yards away, and then watched as I drew nearer and they became larger. Try as I might to recognise the nature of my delusion, the reality that underlay them, I could not do so, and once or twice I actually walked close by my fictional beings, peering at them from every angle as I approached, passed and left them behind me. It occurred to me that here we might be experiencing something quite traditional. We were on the lands of native North American Indians, and many of them had used periods of fasting and wandering in the wilderness as part of the initiation into adulthood. They believed that, in this way, young men would meet their Gods, and now I wondered if I had met them too. Many of the spectral figures formed from trees were giant living totems.

My father still claimed that he was not seeing illusions – a denial we all accepted until he stopped to have a pee. As he returned to pick up his bike, we were astonished to see him drop suddenly to his knees.

'Look,' he said, his voice filled with wonder.

We all looked but could see nothing except his bent figure crawling forward as he brushed his fingers back and forth across the rocky surface of the mountain track.

'It's carpeted,' he went on. 'Beautiful . . . Why have they done that?'

Clearly he had not had enough sleep after all.

Sometimes we had identical visions. At one point my father, now hallucinating freely, and I were baffled to see the others standing in the road amid piles of office equipment. We could even point out imaginary items to one another. Yet, when we drew nearer, the imaginary world dissolved – desks, chairs and tables melting back into tree trunks, branches and shafts of moonlight.

A little after midnight we reached the top and crossed the pass. Now we could benefit from all the height we had gained, for down the other side would be miles of non-stop descent. Getting on the bikes and free wheeling through the darkness was a welcome respite, but it was not easy to see where we were going on the darkened road, especially with moonlight and our brains continuing to play tricks. We also had to be careful to avoid branches and sharp stones on the track. Unable to steer with accuracy, we ran over things that we might normally have avoided, and it was not long before we had our first puncture. I worked on the bike while the others lay down on the road. It only took a few minutes to complete the repair, but when I had finished and called for us to be off, I was startled to find that the others were fast asleep – not just dozing lightly but needing to be shaken in order to rouse them. The same thing happened repeatedly as we descended, stopping to deal with five or six more punctures.

At the foot of the hill our unmade track joined a tarmac road, and for some miles we pedalled along quite happily. The danger now was dropping off while cycling. Relief appeared in the form of a town which we entered in the early hours of the morning, passing blackened shops and offices, a silent railway station and some sleeping residential roads. Then we left it behind and began a long climb back into the mountains, re-joining a high logging road that would carry us on for the next twenty miles.

Our lack of speed uphill was our saviour for, away from the town lights, our spectral visions returned with a vengeance, and while we were cycling they were positively dangerous. Visions would appear on the track with no warning, and when half asleep one would take sudden evasive action. The road was narrow, bounded on one side by rising rock and on the other by an ever increasing drop. A nasty accident could have occurred had we been travelling any faster.

At the top of the hill, the safety factor of slowness disappeared. Having gained enough height to pass above the huge cliffs, the road now contoured along the mountainside, passing over precipices, rising and falling gently to match the terrain. Inevitably we became faster on the downhill sections, and when our efforts slackened off, our lapses into sleep or visions also became more frequent. After only five minutes of this, David at the front stopped and waited. As we approached, he spoke.

'This is ridiculous. I keep swerving to avoid things that aren't there. We have to rest.'

'It'll be okay once it's light,' I urged. 'Surely we can keep going till then.'

'But someone's going to have an accident,' David insisted. 'It's just too dangerous. What do you guys think?' He turned to the others. 'Should we vote on it?'

The others agreed with David and, reluctantly, I had to accept the mutiny. At the same time, I was adamant that we would not stop for long. Although we had been up for 45 hours, I knew from my previous work with the military that a sleep of just forty minutes can restore vigilance in situations such as tired men watching early warning radar screens. Longer naps – unless of many hours – tend to make matters worse. I set my watch alarm for just forty minutes' time and joined the others lying in the road. They were already fast asleep.

Two hours later I was woken by my father shaking me. My watch had failed to rouse me, and what had woken him was the splashing of rain. It was almost 4.30 a.m., and with no time to waste, we got the others up and restarted. The sleep had helped, and after an hour the light also improved, although it revealed a murky sky. The rain that had woken my father came to little, but the cloud over the mountains ahead looked extremely threatening.

Another three hours passed, during which we made fair progress, although the road became a bit of a roller coaster and we were not all able to cycle up the steeper hills. Two or three teams appeared in the distance behind us, but it did not seem to matter for we knew we would now make the next checkpoint. Indeed, it was fun to watch them as they laboured closer. When they caught sight of my father, you could see astonishment written on their faces. Here they were, nearly seven days into an event for which they had given their all, and only now had they caught up with a 70-year-old. If it knocked their egos, it certainly made me smile.

The storm hit us at around 9 a.m., coming quite suddenly. The drizzly rain had been and gone a few times earlier, but now a sudden wind developed and was soon followed by an intense downpour that turned to vicious hail, beating on our clothing. For about half an hour it was unrelenting, then the wind died away and left regular heavy rain which continued for the next few hours as we descended from the mountains into another town. From here it was just twenty miles along a surfaced road all the way to the stage finish. We had as good as completed the ride and began to wonder what was next.

An hour later, while pedalling flat out, a sudden movement on the other side of the road caught my eye. It was a large white dog, coming out from a farm gate, and the sight of my bike seemed to arouse its ire. Barking furiously, it ran straight for me as I swerved. Although I got past the dog, at the speed I was going I could not regain control and was soon off the edge of the tarmac and on loose shingle. This removed any remaining chance of steering and, still going at great speed, I left that as well and plunged down a ten-foot rocky ditch.

For reasons that have never been clear to me, I took the whole impact on my right shoulder, leaving everything else unharmed. The shoulder pain was so intense that I retched repeatedly. After a few seconds, the waves of nausea settled and I could think more clearly. I was sure that it was broken. Slowly, however, as the initial shock subsided, I realised that I could move it a little, and that made a fracture less likely. I had been fortunate, and as the others caught up and halted anxiously, I reassured them that we could carry on.

The reassurance was fine but pretty optimistic. My shoulder throbbed terribly. If the white-water rafting came next, how could I steer? The weather deteriorated further, rain falling from leaden skies, driven by the wind, and we were utterly sodden and cold by the time we turned from the roads on to a track leading to the finish. The track was a sea of mud, churned by the race support vehicles which had preceded us. In fact, those vehicles had been that way twice, for we had returned to the very field by the Lillooet from which we had started. More than a week of travel had seen us complete an enormous circle through the mountains, and now we would have to repack quickly and be off again.

Coming into the field, all was strangely quiet. A couple of teams that had passed us earlier in the day were crouched by their tents. They stood up to clap and whistle as my father appeared. But where was everybody else? The majority of teams were still behind us, and so most of the support crews should have been here and waiting. We caught sight of our own vehicle, close to the river and the administration tent. Mo and Moira were nearby but, although they too waved and clapped as we approached, they had an air of dejection about them and wore frowns instead of smiles.

'What's happened?' I asked.

'It's off,' said Mo gently. 'Nobody's going any further.'

'Why?' I demanded. 'There's plenty of time yet.'

'The storm,' she replied. 'It's hit the teams in the mountains and they're all lost or stuck. They won't let anybody else up there.'

Slowly, with many interruptions, the details emerged. The rain and hail which we had encountered that morning had caused havoc in the mountains ahead where the next stage was a short but technical climbing section. A foot of snow, driven by 70-mile-an-hour winds, had pinned down all but the leading three teams who had already crossed the highest ground when it hit. Tents had been destroyed, and big crevasses on the glaciers had become extremely dangerous. A rescue operation was in hand to lift the teams off the high ice before another storm struck. The weather forecast remained appalling.

At first, I was furious. I had not come so far to be told that I had to stop because it was breezy and snowing. Ran and I had crossed the largest ice field on the planet in weather that made this look like a summer's afternoon. We were not going to be dictated to about what we could or could not do here. I raced up to the administration tent and started to argue. My impression was that the storm was not their only problem. The whole race had been badly misjudged in terms of difficulty, so that teams were now scattered over hundreds of miles. The reason for the absence of many support vehicles was that some teams had not even started the biking section and several, including Helen Klein's, had still not finished the big hike. It all struck me as highly incompetent, and I'm afraid the poor junior staff at the checkpoint got an earful from me.

Slowly my wrath subsided and I began to see sense. If we had come this far, moving relatively slowly, the teams that were a long way back should never have entered the race. Ahead, the snow was wet and heavy and, drifting with the winds, would make deceptive crevasse bridges. Avalanches would also be a threat, and I had to accept that it would be wildly irresponsible to allow any more teams, some of them with little true mountain experience, into the area. If they had to stop some, they had to stop all. I tried to persuade them to let us cycle round the range by road, so that we might still do the white-water rafting the next day. The ride would be long and my shoulder would probably make the rafting impossible but, anyway, they would not hear of it. Their resources were far too stretched.

There was nothing more to be done; we simply had to look on the bright side. Of the twenty-five teams that had gone faster than we had, only fourteen had left this camp before the storm struck. Eleven of those were now being rescued, and only three of more than eighty teams that had started were to finish the whole course. We had done well.

I looked across at my father and saw an exhausted man in pain from those dreadful sores he had carried from the first day. His eyes were sunk in pits of sheer fatigue, yet from within them came tremendous sparkle. He had done what for most, whatever their age, would have seemed impossible, and now he was probably grateful that we had to give up. Yet I also knew that if circumstances had been different, his pack would have been on again in the hour and we would have departed once more. I felt enormous pride. He may have been an elderly weekend walker, but he had taken up horse-riding, canoeing, rope climbing and glacier crossing, and he had kept on going to pass through a stern test of fitness. Above all, he had shown that action was not only the preserve of the young or the super-normal. I dearly hope that I may do the same when I reach his age.

CONCLUSION

★

In Sickness or In Health

Positive health requires a knowledge of man's primary constitution and of the powers of various foods, both those natural and those resulting from human skill. But eating alone is not enough for health. There must be exercise of which the effects must likewise be known. If there is any deficiency in food or exercise, the body will fall sick.

Hippocrates – 5th century BC

ALTHOUGH at times I have strayed far from the theme, this book has tried to examine human performance and health in the context of our evolutionary design. I have argued that the whole balance of our physiology is set up to overcome the rigours of a mobile life with a variable, mainly vegetarian food supply – the life led by our ancestors, who were genetically almost identical to ourselves, until just 10,000 years ago. As a consequence, the resilience of the human frame is far greater than many people realise. This particularly applies to endurance. Although few are blessed with the muscle fibre mix, hearts and lungs to become champions, most could run a marathon if they wished. Indeed, they could go much further, and I have no doubt that many could run over the Sahara or walk across Antarctica if the opportunity arose. I have also suggested that age is less of a barrier to physical fitness than is generally accepted, and indeed it is in later years that the benefits of a continued active lifestyle are felt most. Exercise can truly help to maintain youth but, unfortunately, most people go out of their way to avoid it.

Our evolutionary heritage has also granted us amazing attributes when it comes to environmental hardship. Because the ancestors of all modern humans were still living in the heat of Africa as little as 100,000 years ago and most evolutionary adaptations move very slowly, we all have an extraordinary capacity to cope with the heat if given some days in which to acclimatise and a reasonable water supply. Simultaneously, despite the fact that it has been only ten thousand years or so since some races came to dwell in much colder climes – inadequate time for evolving better physiological defences against the cold – our intellect allows us to operate in the most frigid regions of our planet. Yet the idea that normal members of the population today should ever be exposed to extremes beyond those met while sunbathing or starting the car on a winter's morning, is anathema. We hear of these capacities in tales of people who either deliberately test themselves through sport or expeditions to the wild and remote landscapes, or who meet genuinely life-threatening situations through accident or misadventure. Everyone who hears of them is duly impressed, but they fail to realise that they too have the capacity to survive.

There is also the down side. Although natural selection honed our metabolism to deal perfectly with our ancestral diet, the development of farming profoundly changed the nature of what we ate. At the same time, the advent of civilisation allowed people to pursue less active lifestyles, a change which has accelerated greatly since the Industrial Revolution and even faster in the age of silicon. Now the majority of adults and children in the developed world are increasingly sedentary and eat foods that are unhealthy. The consequence is an epidemic of ill health. A surge in the numbers of deaths from heart disease, strokes and some cancers in the past century has almost kept pace with the decline in mortality from infectious diseases and some other illnesses that has come about through medical and social advance. Furthermore, in the modern industrial world more than half of all adults are overweight and, if current trends persist, half of those will qualify as medically obese by the year 2005. This creates a burden upon health and leads to great physical and financial cost.

So, what can be done about it? It seems to me that, for any individual, there is a clear choice: either one can hope it is not true, that it will not apply to oneself, and carry on regardless; or one can recognise the problem and try to rectify it as best one can within the constraints of modern society. With modern medicine, you probably stand a good chance of reaching old age either way, but your choice may still have considerable influence on whether you live in sickness or in health. It is all a question of risk.

Although a lazy lifestyle with no dietary constraints has wide appeal, it is a route that leads directly to the possibility of getting too fat, becoming infirm, or of experiencing the crushing chest pain of a heart attack. Nobody would be surprised if the engine of a Formula 1 car were to clog and break down after being left ticking over for most of the time with just an occasional run to the supermarket. The same applies to humans. We too are set up for high performance and will clog if we are idle for too long. Furthermore, because of that inbuilt propensity to eat as much food as we can get hold of, particularly fat, we are not just idling but have filled the tank with heavy oil instead of light high-octane fuel. It puts us on to the road to sickness and may cause far more trouble than a breakdown. In our case, blockages can literally affect our lifeblood.

Where, then, is the road to health? The most obvious answer – a return to the lifestyle of our evolutionary roots – would be both impractical and unwelcome. There is no possibility that current society will go away and very few of us have any sane wish to return to the hunter-gathering of our Cro-Magnon past. If we did so, not only would we forego most recreational pleasures and comforts but many benefits to health reaped by technology would disappear. Nobody would be fat or die from coronary heart disease, but millions would perish from starvation, infections and injuries. It would be a truly retrograde step and far more people would suffer. Instead, what is needed is compromise, changes that can be incorporated into our modern lifestyles without imposing too many dietary restrictions or activity levels that are either too unpleasant or too time consuming.

On matters of diet, we could arguably get very close to the ideal. It is entirely practical to consume meat rarely, eat fish fairly often, and otherwise consume a predominantly vegetarian diet. Evolution, however, has given us such a strong liking for meat that many, myself included, would find it hard to cope with eating it only once or twice a week. The compromise for many could perhaps be having meat on alternate days in reduced amounts, while avoiding too much fat from other sources, especially the frying of foods, which turns inherently healthy products such as potatoes into such health hazards as chips. Fortunately, these changes need not solely rely on an act of willpower. The increased proportion of carbohydrate and protein in your resulting diet will allow your control systems to work properly. Your body will recognise much more of your intake and your appetite will be better satisfied accordingly. You will also get a more plentiful supply of micro-nutrients to help with other aspects of health, as well as eliminating any question of hunger being driven by a shortage of vitamins or trace elements. Eat at least five portions of fruit or vegetables every day, include some fish in your diet twice a week if you can, and avoid too much salt and sugar – best achieved by steering clear of too many manufactured foods – and you should be well on your way to health.

Would this be enough? Unfortunately, the answer is no. As was realised by Hippocrates, we cannot maximise our chances of avoiding illness and infirmity if we remain inactive. Even the most healthy of diets may provide inadequate vitamins if you are eating very little because your demands from activity are low. Exercise also brings many other benefits unrelated to food intake, including increased strength of heart, lungs, muscles and bones, but it requires time and effort. Both men and women find it hard to make either.

As a first step, more activity needs to be incorporated into everyday tasks. Unless you are living or working in a tower block, you should avoid lifts as a matter of principle, and when going short distances, you should always walk. Every day in our larger cities, huge numbers of commuters and shoppers take subways and buses for just a stop or two when it is often easier, quicker and more interesting to stroll down the street. Don't always sit. Many jobs may tie you to a desk or screen for much of the day, but informal discussions with one or two colleagues can be made while standing, and the same goes for telephone conversations. Put your phone out of reach of a comfortable chair. Never mind if people think you odd, you can add considerable physical exertion to the day with no cut in efficiency.

All these measures will help your overall energy expenditure and calorie control, but they will still not be enough. For optimal health, you need to stress the system properly. Those of us with muscles close to a fifty-fifty fast/slow fibre mix and average hearts and lungs need surprisingly little vigorous exercise – by which I mean work at a level hard enough for the heart rate to rise and make you short of breath – to achieve 90 per cent of our genetic fitness potential in terms of aerobic capacity. Just one hour a week of aerobic work will see most people up to this level. It would see your oxygen delivery systems ready for a marathon run, even if it would not be quite enough to make muscles, bones and joints ready for such a race. If you finished, you would suffer terribly. Of course, many might feel that such a consideration is irrelevant for they have no plans to run in marathons anyway. But only by exercising frequently can you develop muscles that will clear excess fats from the circulation, produce platelets which are less sticky, bones strong enough not to break, and joints lined with tough cartilage that cannot easily be harmed. To do that will take two or three hours of hard work a week and, if you are anything like me, it can be hard to make yourself do it.

Why is exercise so hard to undertake? Boredom, discomfort, fatigue and lack of time must all be contributory, but there is an additional problem. Goals such as good health in old age are far too nebulous to provide the motivation. For that reason, it is often only those with a more definite short-term goal that succeed. People who need to lose weight, for example, are more likely to continue to put in time and effort than those who are not too fat. If you give yourself a definite aim, exercise acquires a purpose. You need to set yourself a challenge.

What sort of challenge is required? The answer must be different for every individual and its nature is probably unimportant. What does matter, however, is that you seek to undertake something physically demanding that is beyond your currently perceived limits. Rather than simply taking up jogging, you should decide to do next year's local half marathon. If you already run, or a half marathon looks too easy, decide to attempt a full length race in London, Boston or New York. Of course, it need not just be running or an organised event. Plan to spend a weekend mountain biking along the full 80 miles of the South Downs Way or decide to walk around the coastal path of Cornwall. It little matters exactly what is performed, so long as the undertaking is difficult enough that achieving it will bring pride and self respect. Take on something hard enough and achieve your goal, and you will never regret the experience. The capacity to overcome hardship is inherent in us all and, if we use it, we not only gain health but find satisfaction.

And what for me? Will I go on looking for challenges in life, or settle down and hope that what I have done up to now will protect me from the ravages of heart disease or osteoporosis? Will it be enough to allow me to live to a healthy old age? I think not. Indeed, as I write these final lines another challenge has appeared on the horizon. Just three weeks ago, another quiet Sunday afternoon was disturbed by the ringing of the telephone. When I heard the soft American drawl of Mary Gadams, I thought she must be calling to suggest another go at the Eco-Challenge. This year's race – to be held in the Atlas mountains and the Sahara – had just been announced. No, she had a quite different proposal.

'Mike,' she said. 'Are you still running?'

I assured her that I was, though the writing of this book has curtailed it of late.

'Would you be interested in joining in some marathons for charity?' she went on.

There was something in the 'some marathons' that sounded suspicious but I responded positively.

'Sure,' I said. 'How many have you in mind?'

'Oh . . . seven,' she replied, but there was teasing in her voice. 'One in each continent.'

POSTSCRIPT

★

Running the World
Seven Marathons, Seven Days, Seven Continents

Bristol, early June 2003 – Ranulph Fiennes, my long-time partner from Polar expeditions and the British Columbia Eco-Challenge, had just boarded a flight for Edinburgh when his heart stopped. Unbeknown to him, life-threatening deposits of fatty chalk had developed inside his coronary arteries, the blood vessels that supply blood to the heart muscle itself. Now, one of these deposits of atheroma had split and the raw internal surface that this exposed triggered his body's clotting system. Unfortunately, this did more harm than good. Designed to cover and repair breaks in blood-vessel walls, the clotting system is great for cuts and injuries, not so good for heart attacks. The clot soon caused further narrowing of the artery and it blocked. Downstream, a limited area of beating heart was suddenly deprived of life-giving oxygen and its regular, coordinated beat degenerated into chaotic squirming. Of itself, the loss of that small part of his heart muscle might not have been so serious, but the random electrical signals generated by the dying cells soon spread widely and the function of his heart was totally disrupted. Ran was in ventricular fibrillation and had only moments to live.

At this point, Ran was lucky. Just across the runway was a fire crew trained and equipped for emergency defibrillation. Within minutes, the aircrew's mouth-to-mouth life support was superseded by the arrival of a two-hundred-joule DC shock, which passed right across Ran's chest to depolarise every cell in his heart. The hope was that his heart's pacemaker would recover first and it did. His heart was once more a regularly beating pump. Deeply unconscious, Ran was rushed to a waiting ambulance.

Twice more on the journey to Bristol's Royal Infirmary and three more times in the A & E department, Ran lapsed back into fibrillation.

The stabilising drugs he was given proved ineffective and, in desperation, the doctors sent him to surgery. Within just two hours of his initial collapse, Ran was on cardiac bypass with a machine artificially pumping and refrigerating his blood. Meanwhile, his heart, now deliberately stopped, was subject to the finest of surgical skills and the narrowings in two of his three main coronary arteries were bypassed using short segments of vein stripped from his lower leg. Finally, at the end of the three-hour operation, Ran's heart with its new plumbing was successfully restarted. Weakly at first but then more vigorously, it took back the job of circulating his blood. The bold intervention was over and Ran was moved to intensive care.

*

Why had Ran, a supremely fit and active man, suffered a heart attack at such a young age? Any answer is of course speculative, but the issue boils down to a question of risk vs. protective factors.

In Chapter Eight, I described how the adoption of sedentary lifestyles, fat-filled dietary choices and smoking had made heart attacks the number-one killer in the Western world, proving fatal three times more often than breast cancer in women and four times more often than lung cancer in men. Ran smoked enthusiastically earlier in life and never paid much attention to 'healthy' aspects of his diet. Furthermore, he probably has an inherited metabolic predisposition for heart problems. In many, this would have been recognised from their family history – a mother or father having a heart attack at an early age. In Ran's case, however, this could have been missed. He knew his mother had a healthy heart to great age but his father's medical potential was unknown (he died young in the Second World War). Ran's cumulative risk factors simply overcame even his level of physical-fitness protection.

Of course, the newspapers had a field day. Although few doctors make the causal link between ill health and our bodies' evolutionary expectations, many have regaled the public with suggestions on healthy living for decades. This advice has been largely unwelcome. The last thing that the overweight and/or lazy wish to hear is that they should not be sitting about eating, drinking and smoking. There is therefore an inevitable counter-press who identify any story that might justify unhealthy living. Just as the tobacco industry used the occasional tales (*the forty cigarettes a day washed down by a bottle of whisky man living to great age*) to argue that smoking had no relationship to lung cancer, many journalists are more than happy to promote anything that makes activity sound harmful. Their pieces range from those with a modicum of truth (*the experts keep changing their minds and hence we can't be sure of anything*), through the apparently plausible but still ridiculous (*exercise is bad for your joints and you will all end up with arthritis*), to the ludicrous (*exercise uses up too many heartbeats*). For this school of journalism, the story of Ran's heart attack was of great appeal. Sir Ranulph Fiennes, the epitome of continued vigorous activity in middle age, had clearly been overdoing it. It annoyed me intensely.

Just as we know for sure that smoking kills through cancer, there is no doubt whatsoever that regular exercise protects you against heart problems. In fact, lack of fitness is a far bigger risk factor than the much more widely cited obesity. If you compare the risk of a lean active individual dying of a heart attack to that of those who are overweight yet fit, the increase is only 20 per cent in the latter. In contrast, unfit individuals even when thin are more than 300 per cent more likely to die from heart problems, and those who are both fat and inactive almost 400 per cent more likely. Furthermore, if you are fit but unlucky enough to have a heart attack, you will recover much more effectively. A point that Ran was just about to prove.

*

I first heard of Ran's crisis when I received a phone call in the middle of the night from Dr David Smith, a cardiologist working in Exeter and another of our British Columbia team-mates. The news did not sound good. Even several hours after surgery, Ran remained unconscious, for the brain is very susceptible to a cut in the blood supply that brings it oxygen. Although there was now a reasonable chance that Ran would survive, it seemed inevitable that he would suffer cognitive damage and it seemed likely that he would also have long-term physical disability.

I immediately rang the hospital intensive-care unit to send best wishes to Ran and his wife Ginny. She must have been incredibly distressed, and she now faced a future without the man who had matured from childhood sweetheart into a husband and long-standing expedition partner. But although concern for them both was upper-most in my mind, I have to admit that, for me, Ran's condition also raised practical issues.

Only a few months earlier, at the beginning of January, Ran had called me out of the blue. It had been ten years since we had crossed Antarctica together and six since our British Columbia Eco-Challenge. We had rather lost touch and Ran thought it was about time that we tried something new. Being Ran and aiming high, his first thought was an ascent of Everest, but although this had appeal, it was highly impractical. To climb the world's highest mountain entails a departure in early spring followed by months of slow, steady, staged ascent in order to acclimatise to altitude. My busy job as a hospital consultant and my family made this impossible. I needed a shorter, sharper alternative. I knew of just the thing.

In the previous six years I had done little in the way of big, physically testing challenges. This was not for want of trying. After Mary Gadams had rung with her idea to run seven marathons, in seven days, on seven continents, planning had commenced immediately with the event scheduled for the following year, 1998. Then, with arrangements quite advanced, it was suddenly deferred for a further year. A year later, it was deferred once more and with the second delay came a new demand. Each race participant had to come up with $200,000 in order to take part. I was forced to withdraw, somewhat embittered by the waste of three years in which I could have been involved in events that actually happened.

At that time, I assumed that the seven-marathons challenge would go ahead without me but Mary and her US team never made it happen. The project had been dropped. Now it seemed the perfect idea for Ran and me. I felt a little guilty about 'lifting' the concept but not so concerned as to forgo it. Even though Mary could not pull it off, perhaps we could.

Ran needed no second invitation. From that moment he put his heart and soul into the organisation and, by May 2003, everything was falling into place. Both he and I had started early training and we were well on the way to securing financial sponsorship from Land-Rover, with British Airways kindly agreeing to provide us with flights. We had also identified a series of suitable venues on each continent. The only major decision outstanding was the choice of the charity which would benefit from our efforts.

The seven runs had to start in Antarctica. The travel in and out of that ice-bound continent is unpredictable and air services are extremely limited: storms blow in at a moment's notice, often precluding aircraft landings. Fortunately, take-offs *can* usually be achieved. This meant that if we were on the ground and waiting, we could start our first run at exactly the right moment, finishing it in time to leave the continent to join the global flight network. The most difficult bit was reaching the Antarctic in the first place and deciding when to start.

The timing was defined by the choice of our seventh running venue. The ideal seven-marathons challenge would comprise seven different official runs in the same week, but this was clearly impossible. I knew of one individual who *had* managed official races on each continent but it had taken him ninety-five days. We therefore had to organise most of the runs ourselves. Nevertheless, it would be nice to finish with a large official race. The obvious choice was New York.

Access to Antarctica is feasible only during the southern summer, between the end of October and January, and the New York Marathon is run in early November. 'Exactly the right moment' for the start of a seven-day challenge was therefore 7 × 24 hours before crossing the finishing line of that race. Whatever our physical condition, we thought that this would be within seven hours of the start. Since that would be at 10 a.m. on 2 November we could set the time to commence our Antarctic marathon by counting back seven complete days from 5 p.m. on the same day. It brought us to the early evening of 26 October 2003.

★

With my long-time partner in a coma in intensive care, all the thought and planning that had gone into the seven marathons seemed to have come to nought. This did not, however, reckon with Ran's resilience. On only the fourth day after his surgery, he rang.

'Mike?'

I immediately recognised his voice.

'Ran, how are you? How's Ginny? How's –' My flurry of surprise was swiftly interrupted.

'Mike, I can't chat for long but I'm fine. Don't cancel anything for now.'

I guessed immediately what he was referring to but I needed to be absolutely certain.

'You mean the marathons?'

'Yes,' he replied. 'I've been told there was little damage so I don't see –'

'What? You mean your doctors say it's okay?' I interrupted.

'No, no,' he responded whispering. 'I haven't asked them yet.'

Our conversation ran on for a couple more minutes but at the end I was clear. Ran was in cloud-cuckoo-land and his specialist would soon bring him down to earth.

Then five days later he rang again.

'How's it going? When will you be home?' I started.

'Oh very soon,' he came back, a slight laugh in his voice. 'I'm just walking off the moor.'

This was no normal approach to cardiac surgery rehab. I needed to give things further thought.

Could Ran possibly manage the challenge after all that had happened? Although the exercise would clearly stress his heart, the intensity of our proposed runs was never going to be excessive in cardiovascular terms. Our aim was simply to finish – to maintain the slowest possible pace in the hope that this would see us through each event in good enough condition to start the next. Our friend and heart specialist David Smith raised no major objections and this was doubly valuable since David had seen the angiograms from the time of Ran's initial heart attack. Angiography uses high-speed X-ray filming and dye to outline both heart function and the state of coronary blood vessels. In Ran they demonstrated that the damage to his heart was limited to a small area and that his surgery would have bypassed the major narrowings in two of the three main blood vessels. A slight stricture that was left in the third main coronary vessel did not look likely to give trouble. Overall, his heart was a good strong pump.

Our project was back on the rails and we now had the perfect choice of charity. The British Heart Foundation strongly supports the place-ment of out-of-hospital defibrillators of the type which had saved Ran's life. We therefore decided to raise money for them. We also decided to take one with us.

*

I have never been much good at maintaining fitness without a challenge to force me out and on to the hills. The trouble now was that achieving the sort of fitness we required took so much time. Even before Ran's heart attack, I had found that job and family commitments constantly interfered with my plans to run. I did manage perhaps a half-marathon on a Sunday with a couple of shorter runs elsewhere in the week, but this never met my aim of completing several big distances a week so that my body would become used to facing long outings in quick succession. Furthermore, when in May I had tried a full 'official' route – the so-called 'Neolithic' Marathon from Avebury cross-country to Stonehenge – I was unable to finish without walking. My self-diagnosed fitness report read 'F – must try harder'. Then, through June and much of July when Ran's heart condition made the whole challenge unlikely, I had gone back to just an occasional short run. Now the show was on again and it was nearly August. There was little more than three months to go.

Ran too was far from ready but his training problems were different.

Although he had the time, he had been told to take things easy. The surgeons were keen that their carefully sutured bypass vessels were not put under too much pressure, and Ran was allowed only a carefully building exercise programme in which he walked further and further each day. Clearly, since he would not be permitted to run at all for a few more weeks, this was far from ideal in terms of 'fitness for task' but, nevertheless, his ever brisker walking over the Exmoor hills must have induced the adaptations of endurance fitness described in Chapter Four. Slowly but surely his heart pumped more blood per beat, his lungs shifted more air per breath and his muscles powered through more and more steps with greater strength and less fatigue. Tendons, cartilage, joints and bones also strengthened. Steadily he became more capable of facing the punishment ahead.

Decisions on the venues for the runs between Antarctica and New York had been dictated by flight logistics and opportunities for media and charity interest. A number of cities had been considered but, as the challenge drew nearer, we finalised on Santiago de Chile for South America, Sydney for Australasia, Singapore for Asia, London for Europe and Cairo for Africa. As each decision was made, local organisers were identified to deal with the exact choice of marathon route and other practical arrangements. This produced interestingly varied attention to detail. Some of the organisers simply suggested a course of roughly the required distance and took a 'we'll check it on the day' attitude. Conversely, others became obsessed by 'official' measurements, with long e-mails discussing verification techniques and the allowance that should be made for the correct 'tangents' that we might run on any bends. To my mind, the latter were wasting their time. As long as we completed at least a marathon (and we were perfectly happy to a run an extra couple of hundred yards), we did not care. Official finishing times were of little interest.

With less than three months to go, there was still too little time in my day to work, train and see my family. I therefore switched to a somewhat risky approach. Instead of a steadily increasing running programme, I decided to push myself really hard by running a marathon or more every Sunday while fitting in shorter runs when possible. To ensure this, I entered 'official' events in Chichester, the New Forest, Winchester, Cardiff and Richmond and, in between, created my own courses through the local countryside. I found to my delight that marathons did become easier. By the end of August, I had gone from being barely able to walk for several days after each outing to being comfortable and walking reasonably by the following morning.

Nevertheless, my legs were still utterly drained by each event and the prospect of running another marathon within days let alone hours seemed distant. But I was not too disheartened, until disaster struck.

Long outings, even for those in peak condition, stress muscle, bone and tendon. I knew from past experience how easy it is to get injured. My main fear when it came to the seven marathons was of overuse injuries: tendonitis, blisters, a swollen knee, even a stress fracture. To give my body less of a pounding, I trained 'off road': although it may be easier to turn an ankle or twist a knee on a cross-country trail, the ground is softer and your stride more varied. But one Sunday, eight weeks before departure, I was out on a long run when my left hip began to complain. I had already covered more than a marathon distance and was just a few miles from home, but although I tried to ignore it, within minutes I was in agony. I was forced to call my wife to pick me up by car. Back at the house I crawled upstairs for a shower. Over the next few hours, a grapefruit-sized swelling appeared which slowly became a massive, blue–black bruise tracking down my thigh. I did not know what was wrong but I was worried.

As a hospital doctor, it was easy for me to get an ultrasound the following day. The results were not encouraging. A small muscle – the tensor fascia lata, which runs from the hip to beyond the knee – had torn completely. The ripped ends were swimming in a pool of blood. My orthopaedic and sports-medicine colleagues were universally gloomy: I would not be running anywhere for several months, let alone embarking on seven marathons. Fortunately, I didn't believe them. The injured muscle, in fact, contributes only slightly to stability at the hip and knee and so one can do without it. More importantly, although partial muscle tears are very prone to reinjury, in this case there was complete division. As long as the torn ends did not bleed again, I did not think it would give me trouble. I took two weeks off, several aspirin and had some physio before a tentative half-hour outing. The hip was sore but became no worse. The following weekend, three weeks after the pronouncements of doom and five weeks before our challenge, I joined Ran for our first training session together, the Cardiff Marathon.

A few weeks earlier, when Ran had opened his posted Cardiff registration details, out had popped the race number 777. He immediately assumed that I had arranged it. This was untrue but surely it could not be coincidence. Perhaps the event organisers had heard of our plans and recognised Ran's name on the entry form? When I rang them, they had no idea what I was talking about. The number had been allocated by chance. It occurred to me then that I too should have a special number and the Cardiff organisers were willing to oblige. On the morning of the event, Ran wearing 777 was surprised to see my 999. The irony of the choice was not lost on him. This was his first full-length run since his heart attack. For me too, it was also make or break.

The Cardiff course involves two loops of the city and bay – a route that is flat, attractive and easy. The sun shone and by halfway I was confident that my hip was fine. Nevertheless, my enforced three-week lay-off had sent my fitness backwards and during the second loop I was tiring, even though we took things slowly. We eventually finished in around four hours. Afterwards, sitting on the steps of City Hall, I felt pretty dismayed. It seemed that my endurance fitness had regressed by several weeks. My only consolation was that we planned to run the seven marathons at an even slower pace – perhaps five to five and a half hours – in order maintain our strength.

In contrast to me, Ran looked pretty fresh at the end of the Cardiff run despite the ten years' age difference and his enforced, slow training schedule. It was easy to see that he was by far the more natural athlete. But he too had a problem. For many years, his back had been the source of much discomfort, causing considerable grief when he was pulling heavy sledges during some of our Polar trips. Sadly running also made it worse and if one marathon exacerbated the pain, how would it react to the hammering it was going to receive? We were both aware that determined or not, enough pain could defeat us. I made a mental note to add more painkillers to the medical kit.

*

Having recovered from Cardiff and the Clarendon Way Marathon near Winchester the week after, Ran and I finally set off for Antarctica and the start of our challenge on 21 October. With us went a small media group: Giles Whittell and Gill Allen from *The Times* and Julie Ritson and Rob Hall from BBC News. Our eventual destination was King George Island, which lies just off the tip of the Antarctic Peninsula. This huge finger of mountains extends northwards from the main mass of Antarctica towards the tip of South America, and it is by far the most popular place for different nations to establish research bases. King George V Island has several, served by an airstrip run by the Chilean Air Force. This made it the perfect location for the first leg of our challenge – only five days away now – but getting there would not be easy. First stop was Punta Arenas, close to Chile's southern tip and four hours' flying time from Antarctica.

We arrived in Punta nearly thirty hours after we had set off, our first taste of the enormous flights which our challenge entailed. Fortunately, thanks to British Airways we would spend most of our time flying First Class, and this meant both better food and more comfort. But despite the roomy reclining seats on our outbound journey, Ran's back had already become worse. He was trying to hide the pain but it was written in his face. In fact, his back had deteriorated before we had left the UK. He had called to warn me a couple of days before departure, even talking of potential failure, which was quite unlike him. Now, his prospects looked even worse.

For the next two days, the weather in Antarctica was appalling and I was glad we had given ourselves some leeway. Then, although the situation improved, we met an unexpected problem. The bad weather in Antarctica had preceded our arrival in Patagonia and now there was a backlog in vital Chilean Air Force resupply missions. Argue as we might, there was no chance that we would be allowed to fly in before or during these operations. The Antarctic airstrip was too small for another aircraft to land when one of their Hercules transporters was on the runway. Deeply frustrated we spent several days continuing our training along windswept beaches out of town, the Magellanic penguins on the shore providing a reminder of just how close we were. But when it comes to the Antarctic, close is not enough.

On 25 October, the day before we had to start, we received our first good news. The Chilean Air Force had completed their resupply flights and a short but promising weather window was coming in. If we flew that day, there was a good chance we could land, spend one night on the ice and then start running the following afternoon. As we flew south, crossing the glaciers and mountains of Tierra del Fuego, our excitement mounted. It was soon to be dashed. After less than two hours, the captain announced that the weather was closing in ahead of us and that we would have to turn round. But back in Punta the situation looked less bleak. Our meteorologist was in a much more positive frame of mind. A satellite picture showed a huge area of fine weather approaching the Peninsula and he was confident that we could fly the following morning. This would still give us time to start our seven days of madness.

The weatherman was correct in his confidence and the next morning we set off once more, full of hope. But as we taxied and turned to face up the runway, Rob Hall made a statement that clearly rankled with the gods. Just as he announced that 'nothing can stop us now', the starboard engine coughed and spluttered to a halt. A few moments later, the captain confirmed engine failure. Our seven-continents plan was dashed.

Back in the terminal building, difficult decisions had to be made. We could try to reschedule the whole project – neither Ran nor I was keen on this option. We could go to Greenland after New York – this seemed both difficult and contrived. Perhaps we should abandon the plan completely. Then came the brainwave. Although not part of Antarctica itself, the Falkland Islands had always been the administrative capital of British Antarctic interests. The Islands would make a good proxy for Antarctica if only we could get there in time. This was hugely problematic: we needed to start running in just four hours' time and we no longer had a functioning aircraft. To make matters worse, flying to the Falklands from Punta Arenas entailed travel over Argentinian airspace. Permission was not likely to be granted. Nevertheless, it was our only real hope.

We immediately adopted this new plan with purpose. Ran and I would drive out of town to run our first marathon on the beaches where we had trained beside the Magellan Straits. This, instead of Santiago, would be the South American part of our project. While we ran, our local organiser Alejo Contreras would try to procure an alternative aircraft with not only the range and speed to fly us to the Falklands but the capability to take us on from there right across South America to Santiago, where we could lock back in to the complex web of worldwide flights that we had previously arranged. Alejo would also enlist British diplomatic help in his efforts to sway the Argentinian authorities to permit the necessary over-flights. It was a bold change of scheme that would need a lot of luck to work. As Ran and I changed into running gear in the airport lavatory, it seemed more than likely that Punta would be our one and only run.

My initial idea had been simply to run our 26.2 miles on an unmade track by the coast. But as we drove away from the airport, I noticed that it was extremely windy. This was important and I turned to the map. The coastline changed direction halfway through the route – though we would have the wind at our backs for ten miles or so, later we would head into the teeth of a near gale. Fortunately, there was an alternative. A different track left the road we were on, not far from town. It looked as if it would allow us to run along the back of a different beach, before moving inland across low moors, all in the direction of the prevailing wind. We hurriedly changed our plans and took this untarred turning, zeroing our milometer as we left the main road, since we would be running back to that junction. Then, after twenty-six miles had ticked by, we added a few hundred yards and pulled in. We were at our start point and had just under an hour before that critical calculated moment, seven days before we must finish in New York. A mad idea, months of planning and hours and hours of running had led up to this unique moment. Here we were on a remote beach in southern Patagonia; I had no idea as to the likely outcome.

★

As Ran and I waited by our start line – well, an imaginary line on an unmade road – I had time to worry about the things that might go wrong. I had been so taken up with our Antarctic difficulties that I had put all the other problems that might stop us to the back of my mind.

First, there were the medical issues. Of these, sports injuries remained the greatest threat. Although training had gone well – no tendonitis, blisters, sore knees or stress fractures so far – my hip and Ran's back had supplanted these routine ultra-distance problems. They were now a real risk. Separate to the injury issues, came the possibility of illness. Ran's heart might be unlikely to cause problems but viruses or stomach upsets seemed all too possible. Even one twenty-six-mile race depresses immune function and makes you more vulnerable. Furthermore, our endless air flights, sitting with hundreds of others in enclosed steel tubes, hugely increased the risks of viral infection. Running a marathon with flu-like illnesses is both difficult and risky, for viruses can affect the heart. A number of unexpected deaths in super-fit young Scandinavian orienteers have been attributed to such unrecognised cardiac involvement. On the other hand, the endless flights round the world would limit risks from stomach problems. Although we would graze on available food as we went, all major airlines take great care not to poison their passengers. We would therefore be unlucky to contract dysentery.

There was also another category of medical risk over which we had no control at all: threats from the environment. As I have discussed earlier, heat production when working hard effectively annuls dangers of hypothermia. Risks of hyperthermia, however, were another matter. If we made it through our South American, Falklands and Sydney legs, we would be heading for Singapore. There it would not only be extremely hot but also very humid. In fact, overheating could literally stop us dead. Still, we had to get there first and that brought to mind a second, quite different, type of show-stopper. Our logistic plan was desperately vulnerable.

In order to complete our challenge, we were totally reliant on flights operating to time. If only one was significantly late, everything might fall apart. And once on the ground, we also needed our local arrangements to work perfectly. It would do us no good to reach the airport on time only to find nobody to meet us because of rush-hour traffic. Our timetable was like a complex unstable machine and we were dependent on every cog.

Perhaps most unpredictably, I wondered whether we would fail through simple fatigue. It remains hard to run a marathon even when well trained and this simple fact had remained with me throughout the weeks of preparation. Although evolutionary inheritance grants an enormous capacity for endurance, it is really the ability to take on exercise of low intensity for hour upon hour, day after day. The runs we faced were going to be extremely demanding, probably a lot harder than the Marathon of the Sands. Not only would our total distance be greater but it would largely be spent on roads. Of course, while this makes it easier to run, it also makes the work intensity rise. Hard roads not only cause injuries but also put higher demands on exactly the same muscles with each punishing stride. In the sand dunes, we had often been forced to walk. Finally, if we got further than that run in Singapore, there was the prospect of London and Cairo. How could we cope with our fifth and sixth marathons falling on the same day?

*

Within moments of the start in Patagonia, we were thrilled with our decision to change course. It had been inspired. Gales nearly always blow near Cape Horn and the wind literally pushed us along. For the first few miles we ran beside a storm-tossed sea flecked with spume in the bright, evening sunshine. We were going well and it was soon clear that this first race would be complete in nearer four than five hours. Even ten miles on, when we left the coast to climb towards the moorland, the rising twisting track caused little suffering. The wind remained at our backs and our pace scarcely slackened. Grazing the moors were small groups of rheas, the South American equivalent of the ostrich. Like their African cousins, these birds are basically a long neck and light body perched on two giant muscular legs. Our gait was pathetic beside them as they powered away at perhaps thirty miles an hour.

With sunset came change. We had reached the top of our ascent and the wind now dropped as we went over the crest of the hills. I felt almost serene as we coasted down in the darkness. Amazingly quickly, it was over. We had crossed the line in only three hours and forty-five minutes, and I was elated. Our first test seemed to have gone with relative ease. When I turned to voice this to Ran, however, I saw furrows of pain across his brow and my delight faltered. Ran had stayed just behind my shoulder for nearly the entire time and had certainly not been pressed by my pace. Yet his eyes conveyed real anxiety as he sat down stiffly. He was suffering with his back and there was little I could say to reassure him. Those extra painkillers might well be needed.

As we headed back into Punta Arenas, the news was good. The charter company whose plane had failed had found us a twin-engined executive jet, which would be with us by midnight. Despite the irregularity of such a mission, the Argentinian authorities had given their permission and we would leave for the Falklands at dawn. Finishing the Falklands run in time to make it to Santiago for our onward connections would be tight but not impossible. The seven-marathons challenge remained on. After a meal at 1 a.m., we even went to bed for three hours.

*

The chartered jet was magnificent – a capsule of sheer-leather and wood-bound luxury that whisked us at great speed from Punta Arenas to the Falkland Islands early in their day. As we began our descent we saw that our arrival was being celebrated in style. An RAF Jaguar took up position just feet from our starboard wing and we had a clear view of the pilot's thumbs-up through his canopy. His encouragement lifted our spirits further. Every Falklander knew we were coming and encouragement would be the order of the day.

Once on the ground at Mount Pleasant Military Airbase we had to decide on the best route for our run. We had to be back on board our plane and in the air in a little over seven hours. Since the run might take well over five we did not have long to make up our minds. In the airbase commander's office a giant map covered a whole wall, and with the aid of his local knowledge we made our route choice. The Falklands are very hilly, if not frankly mountainous, and like Patagonia they are very windy, but there was an unmade gravel road all the way from Mount Pleasant to our destination, Port Stanley. On this, the wind would be in our favour, which would hopefully offset the effort needed to climb some of the disturbingly large hills. But where should we actually start? Armed with a reel of cotton we checked off 26.2 miles on the map. Unfortunately, this placed us at the foot of a huge ascent so, after a spot more measuring, we decided to leave from the crest of that hill, giving us a gradual downhill run for the first mile or two. We would then compensate for the lost mileage when we reached Port Stanley by taking a circuit or two around the town – hopefully getting local support along the way – before finishing by the cathedral. An hour after setting down on the runway, Ran and I shook hands with our hastily allocated Falklands helpers and set off at a slow steady lope. We had six hours left before the flight out.

Initially, I felt stiff. The first marathon had gone well but nevertheless, my muscles winced at being asked to work again. Once warmed up, however, things became easier and even comfortable. For the next couple of hours running was a pleasure. We progressed across an unusual upland habitat of heather and tussock, occasionally winding down close to inlets where cold, dark water lapped rocky shores. Our spirits were good and Ran seemed more relaxed. An occasional vehicle swept past with drivers hooting and waving wildly – our progress was apparently being reported on Falklands Radio, and in the capital preparations were in progress for our reception.

One by one, each big rise in the road was conquered and we earned a descent as compensation. Slowly, a distant rocky mountain drew closer. This was Mount Tumbledown, the site of the fiercest fighting during the Falklands War, where British paratroopers who had yomped across the Islands met Argentinian Forces dug in on its flanks. Just beyond, we knew, lay Port Stanley. Every so often our road passed other reminders of the war. Huge swathes of fenced-off land were marked on either side by signs reading 'Slow – Minefields'. Fortunately, we were moving slowly enough.

As with our first run, we had the constant help of our media crew, who by now had become firm friends. They gave us drinks and food, and provided us with vital psychological support. Encouragement matters so much when the going gets tough and tough it certainly became. For me, we were still more than ten miles from Stanley when my legs began to demand remission. They were not too bad on the flat and downhill but on uphills, as we climbed hundreds of feet, they begged me to stop. I had reached the 'wall' and the time for gritted teeth. This is not much fun and I was concerned that I was tiring so early. It did not require genius to work out why.

As I discussed in Chapters Three and Four, the causes of fatigue in activities such as sprinting are well understood – the muscles rapidly run out of their high-phosphate energy sources and lactic acid builds up to poison the system. In endurance running, particularly ultra-endurance, it is far more complex. Yes, the exhaustion of glucose and glycogen does make muscles feel dreadful as the runner hits the 'wall'. But beyond that 'wall' – or for that matter the 'wall' found in lower-intensity activities such as walking huge distances – things are less clear. The cause of this endurance fatigue is probably central – something in our brains that responds to unidentified signals from our body telling us enough is enough. This system, like that of pain, tries to prevent our overdoing things by making us stop before harm can be done. As with pain, however, the signals can be ignored if continuing is important enough.

Our first run in Patagonia would have totally depleted our glycogen stores and between there and the Falklands we had not had time to refuel sufficiently. Accordingly, the 'wall' had moved forwards up the course, and now I could see that with each event to come, it would get closer and closer to the start. Clearly, every run from here on in would be a battle of wills.

But there was some consolation. Both of us had proved our determination repeatedly and I knew well that, if asked, our bodies could give far more than expected. To my mind this is down to our evolutionary heritage. There have always been stories of remarkable feats in life-or-death situations – war or accident making ordinary people push far beyond their limits. The fact that one or more survive is often considered inexplicable, but our evolutionary origins provide rational explanation. Even today, the failure of the rains in Africa may demand superhuman feats from entire populations. This type of natural disaster applied over millions of years of human development and as a result we can all continue beyond fatigue. Nature has given us a 'survival reserve'. Mountaineers, explorers and endurance runners have simply learned to access this reserve voluntarily. If you want to do something enough, you can put your body through hell.

But fatigue was not my only concern. As Port Stanley grew closer, I kept getting twinges of cramp. They occurred every time we climbed a hill and this could be serious. You can choose not to listen when muscles cry 'enough' and it is just a matter of exhaustion, but the body does have other means to make you stop. Cramp in both legs cannot be ignored; it simply takes you down.

After about four hours, we finally crossed a small col near Mount Tumbledown to find ourselves looking down on Port Stanley, now just a few miles away. Here we started to meet groups of people who had driven out to cheer us on. By the time we entered the outskirts of the town, we were at the head of a small procession complete with a police Land-Rover. Despite this police escort, it was here that our challenge almost came to an end. An Islander, ignoring all the vehicles and the flashing lights, almost ran us down. It took a sudden spurt from Ran and me to avoid becoming rare Falkland Islands road-accident statistics.

As we entered the town our spirits rose and our snail-like pace hastened. Ahead, the cathedral with its whale-jaw archway loomed and the end seemed near. But we had reckoned without the necessary route adjustments. To my horror, just as we saw the crowd gathered at the finish, we turned off the coast road and up a hill. This was steep enough to have caused us problems had we been fresh. Although we toiled upwards for only a couple of hundred yards, it was at huge cost. To make matters worse we then headed right out of the far side of Stanley for a mile-long loop before we once again approached the cathedral and the tape. My legs were lead as we crossed the line in four and a half hours to be met by Stanley's schoolchildren, much of its adult population and the Governor of the Islands. He invited us back for tea.

★

Our flight out from Stanley was back in the lap of luxury, although the Governor's residence had already exposed us to how the other half live. Bathrooms with soft towels and robes had been followed by a huge spread of sandwiches, teacakes, muffins and scones. Unfortunately, we had left the food largely untouched: we had had little more than an hour to wash, change, eat, drink and drive for forty minutes back to Mount Pleasant. Back on the executive jet our anxiety levels were reversed. Ran had come through easily – his back problem improved – whereas I, although I had no specific illness or injury, was absolutely wasted. I had found the hills particularly tough and was now becoming more certain that simple fatigue topped my show-stopper league. Perhaps I was paying the price for that missed training. Nevertheless, we had only just started our twenty-six-hour journey to Sydney. Although we would see no bed for a while yet, at least this would allow some time for recovery.

We arrived in Santiago in the early evening, our route extending the length of the day. Nearly all our flights from now on would do the same, for the sun would pass more slowly over our heads as we raced westwards before it. This was deliberate. Our aim was always to finish in Central Park within that 7×24-hour limit, but were something to go wrong with our logistics and we missed a vital connection, our fall-back position was completion within seven geographical days and nights (the dates of our runs looked like eight days but that was simply because we crossed the Date Line). Although it would mean our missing the official New York race, this would grant us nearly twenty-four hours extra time if absolutely necessary. The disadvantage of this plan, of course, was that by travelling continually across time zones our body clocks would become increasingly disorientated.

★

Before embarking from Chile, we sorted through our possessions and reduced everything to one piece of hand baggage each. From now on, making flight connections was the name of the game and we could not afford to wait for luggage carousels. The thirteen-hour transpacific flight to New Zealand departed and arrived on time, but at Auckland airport one of our worst fears was realised. The onward two- to three-hour flight to Sydney was delayed. This was potentially disastrous. Our stay in Sydney was already going to be incredibly short and now there would not be enough time to get from the airport to the course down by the harbour, complete the run and return to the airport for our next connection. In desperation, we contacted British Airways to see if they could organise a later flight to Singapore. We just had to hope that this was possible.

We reached Sydney on the morning of 29 October. Racing through the airport formalities, we were then whisked by car into the centre of the city, interviewed by ABC Radio as we went. Nevertheless, it was only as we were ushered into a huge press conference near the Opera House that we realised we were big news. Australia has an activity culture and we faced a veritable bank of TV cameras and press reporters who almost bullied us for information. Later, when we were out on our marathon, we were recognised by almost every runner we passed – the passing applying only to runners moving in the opposite direction. Sydney is a city of the fit and the runners going our way left us plodding in their wake.

I had always looked forward to the Sydney run. It is a wonderful city and our course had been carefully set up to take us past the sights: the Opera House, through the Botanic Gardens and over the Harbour Bridge. But although the bridge may look beautiful, it caused a problem. The Australian organisers had been a little concerned about the hundreds of steps at each end of the giant structure. Some weeks before we left they had contacted me to ask my opinion about incorporating these ascents and descents. At the time I had had no concerns: I thought that crossing the bridge was no big deal since speed was not the issue. Unfortunately, I had misunderstood the question. What the organisers had done was to arrange a marathon consisting of three repeated circuits of an almost nine-mile course. During *each* of these we would not only have to climb the bridge and descend the far side but also complete a short loop on the opposite foreshore before recrossing. As a result, we actually had to climb and descend the bridge six times. Following the hills of the Falklands the day before, this was to prove shattering.

I was reasonably happy for nearly two-thirds of the distance. We were running with members of the Sydney Striders running club, including the current Australian Iron Man champion, and with their constant quiet encouragement and the best wishes of those joggers out from the office at lunchtime, all was okay until we reached the far descent of the bridge on our second loop. Then a rather unpleasant rumbling began in my abdomen and I realised with dismay that I was going to need to stop and use some facilities. Fortunately, I knew from our first circuit that we were to pass a loo within a few hundred yards and I just avoided a very embarrassing accident.

Emerging relieved a couple of minutes later, I felt a lot happier but my insides were far from right. Since an infection was unlikely, I assumed the disturbance was so-called 'runner's gut'. As a gastro-enterologist, I had seen several cases of this over the last few years. It is thought to be due to minor damage to the gut lining which occurs when too much blood flow is diverted away from the bowel to meet the demands of exercising muscles. I had never suffered from it myself but I had never run three marathons in such quick succession. Minor damage from each was probably cumulative. In the end, it made me stop twice with the runs and this was not the only problem. I began to feel increasingly unwell, and by the time we started on our third and final loop, I was by no means certain that I could continue.

My doubts proved unwarranted and we finished the Sydney marathon in four hours and forty-nine minutes. It was a far from spectacular time and by the end, while Ran remained in reasonable condition, I felt frankly awful. The final circuit had been desperate and although I had managed to maintain a semblance of running, it was with a far from normal action. Furthermore, I felt incredibly nauseated, a symptom I ascribed to a combination of lost fluid and circulating toxic compounds in my bloodstream released by my mildly damaged intestinal lining. The weakness of my legs, however, was more difficult to explain. They were becoming tighter and stiffer in a way that was totally unfamiliar. They were neither aching nor painful; they simply did not want to function. It was as if they belonged to someone else and would no longer obey my brain's commands.

Prior to leaving Sydney, we had one more unforgettable experience. British Airways had succeeded in rearranging our flight times and although our new connection eroded into the rest time we had planned for Singapore, we now had a couple of hours to spare before departure. Our hotel base kindly offered to fill these with a sports massage, something I had never before experienced. At this point, I was prepared to try anything that might help. Somewhat nervously I presented myself to an attractive, slightly built masseuse. Her small frame proved deceptive. She was extremely strong and it felt as if her fingers reached right down to bone. This was both a disturbing and painful experience and in truth I doubt it gave much benefit. Although I had yet to realise it, the damage to my muscles went far beyond excess tension.

★

The first sign that my problems were much more than simple fatigue and stiffness came just before we left the hotel for the airport. Taking a last-minute pee, I was startled to see that as my urine entered the clear water of the bowl it spread like red-brown smoke. I initially thought this must be haemoglobin and that I was suffering from the type of haematuria that occurs in some soldiers when undertaking extended marches. In those cases, it is thought that the repetitive pounding of the soldiers' boots leads to the damage of red blood cells flowing through capillaries in their feet. These then release haemoglobin which passes freely through the filters of the kidney and appears in the urine. When I explained this to Giles, our *Times* journalist, I was not expecting the headlines back home the following day: 'STROUD PEES BLOOD'. By then, I had also made a different diagnosis.

When we arrived in Singapore I was still feeling very unwell; by now I was feverish and had completely lost my appetite. This was very worrying: it was vitally important that I ate. From the very beginning of our planning, we knew that we would have to replenish our glycogen while we travelled, consuming at least a couple of thousand calories of carbohydrate between each run. I had barely managed a couple of hundred. I had also drunk much too little fluid and with my ongoing smoky urine and mild continuing diarrhoea I was becoming concerned that I was dehydrated. As soon as I reached my hotel room I weighed myself.

I had anticipated massive weight loss. It was therefore an enormous shock to find myself weighing six kilos more than when I had started in Patagonia three days previously. There was only one explanation.

The tightness in my legs was actually due to the build up of fluid, and since there was no puffiness around my ankles (where free fluid outside cells accumulates) the excess water had to be in the muscles themselves. That was why they felt so stiff without being painful. But what had caused this fluid accumulation? I had heard of distance runners getting a condition called rhabdomyolysis. This is caused by low-grade damage to muscle cell membranes from mechanical overwork. In such situations fluid from the circulation can leak into the muscle cells while, at the same time, muscle cell content spills into the bloodstream. That muscle content may include the protein myoglobin which supplies the muscle with oxygen (like haemoglobin in the blood). It had to be myoglobin not haemoglobin that I was passing in my urine. I was breaking my muscles down.

This was not a happy state of affairs. If rhabdomyolysis is severe one of the consequences can be large amounts of free myoglobin clogging up the kidneys' filters. Although I had never heard of exercise-induced muscle breakdown at a level to cause renal failure, I had heard of cases in which exercise combined with hyperthermia had done just that. A couple of years before, I had been an expert witness at an inquest into the death of a young, fit Sandhurst officer cadet who had suffered such a fate. And we were about to run in a very hot venue. A big rise in my core temperature might disrupt the already damaged muscle membranes dramatically. It was going to be vital that I stayed cool through the Singapore run – now just hours away. It was also important that I had a blood test to check that I was not developing renal failure. I had no wish for this crazy challenge to cost me my long-term health. Unfortunately, however, that test would have to wait until we had completed this next marathon.

*

By the time we had reached our Singapore hotel rooms, it was well past 2 a.m. All our planning for a good night's sleep had gone out of the window with one delayed flight and since we needed to be up again at 4 a.m. to talk to the press and to prepare for a 6 a.m. start, it was hard to enjoy the now novel sensation of lying on a bed. I was just dropping off into a welcome escapist dream when the phone rang to drag me back to dreadful reality. It was time to force in some food, talk to the cameras and go down to our pre-dawn start.

It was still dark outside the hotel but it was already stiflingly hot. It was also unbelievably muggy and I knew that it was this that would make us suffer. Although we could sweat freely, we would gain little or no benefit. It was simply too humid for sweat to evaporate and the wetting of our skin would not therefore meet its physiological objective. To limit the risk of hyperthermia, we planned to drink continually. We had been using sports drinks to try to delay glycogen depletion and offset dehydration during all of our runs but, as discussed in Chapter Six, these are generally set up with the former as the priority. Here, we were more concerned about maximising fluid absorption than glucose provision and so, as with the Marathon of the Sands, we diluted the drinks to give our gut some chance of meeting our massive losses.

Sydney had had Striders, and here too we were pleased to see that we would not be running alone. A large group of athletes had turned up to accompany us; indeed, the whole event had an air of perfect organisation. This was down to the planning of the Singapore Heart Foundation (SHF) for whom we were to raise money on this leg. They had arranged to send us round the official Singapore Marathon route and although much of this would be through the city's parks, we even had police outriders for the road sections. Hopefully, they would have more luck than their Falklands colleagues when it came to stopping the traffic. The SHF was also useful because of the medical connection. Prior to starting, I asked whether needles and syringes could be waiting at the finish so I could confirm my new suspicion of rhabdomyolysis and check that my kidneys were not in danger.

Within an hour of starting the Singapore run I realised that this was the marathon too far. I felt sick and my legs, although still painless, had become utterly useless as the first few miles went by. The heat was stupefying and I thought my whole body might melt into a pot of tallow. We had managed little more than a quarter of the course before I drew up beside Ran and told him of my predicament. However much I wanted to go on, it was not within my capability and I explained that for me there was no choice but to give in. I urged him to go on running every step if he possibly could. Feeling pretty dismal, I cut back and watched as he and most of our accompanying runners drew slowly away.

I was not planning to give up. From the very beginning of my involvement in the seven-marathons challenge, I had always taken a pragmatic approach. I would run every step of every race if I possibly could. If I could not, I would finish the distance anyway and aim to be fast enough to catch the next flight. If I managed neither, at least I had tried. So, although weak and nauseated (probably from the circulation of my muscle cell contents), I pressed on. Some of the Singaporean runners kindly stayed back to keep me company, and so we commenced a new approach to the route, a mix of alternate brisk walking with running whenever I could manage. Since the running speed I could now attain was little faster than walking, the change did little to slow my overall progress.

The remainder of the Singapore run was hard on both of us. The temperatures climbed and the humidity remained total. Ran too was utterly exhausted: he said later that at one point he had been sure he would fail. Nonetheless, he tottered onwards, drenched with water every few minutes by the helpers who were accompanying him.

Towards the end of our run, the course made a short loop of about three miles with outward and backward paths along the same sector of road. Struggling along, I suddenly realised that Ran was coming the other way. He crossed over and as we approached one another we both raised our hands, which met in passing as the briefest of high-fives. It was a privilege to witness this supreme performance by a man ten years my senior, a man whom so many had recently written off.

Ran went on to finish having run every single step, although he still took five hours and twenty-four minutes. I ended up running about two-thirds of the course and walking the remainder, coming in more than half an hour behind him at just over six hours. I was disappointed, but at least I had done the job.

At the finishing line, syringes were waiting. Ran, despite his hatred of needles, had already given a sample and I was so knackered, I barely noticed the test. I knew I could check my rhabdomyolysis theory by measuring levels of an enzyme called creatine kinase (CK) in my bloodstream. This plays an important role in fuelling muscle function; although present in very high concentrations within muscle tissue, it is not meant to be found in the circulating blood. At the same time, I could also check our kidney function by measuring blood creatinine and urea, tests that are part of any routine biochemical assessment of blood samples.

We were at an SHF charity dinner in our honour by the time the test results came back. I was passed a slim folder by their chief medical officer and I quickly scanned down the printed figures. I was surprised to see that Ran's CK was raised to nearly fifty times normal. He at least was suffering from significant muscle damage. Then I looked at my own results – which took a moment or two to sink in. My CK was a shocking five hundred times normal. It surpassed anything I could have imagined and confirmed beyond doubt that I had suffered massive muscle loss. This explained the trouble with my legs. The only encouraging part of the report was that both Ran and I had normal kidney function and, as an incidental measure, there was no suggestion that Ran's heart muscle was suffering. Still nauseated, I ate little and we had to leave the gala dinner early anyway, heading for our next flight.

*

Boarding the plane from Singapore, I seriously considered giving up. If things got much worse, I would be in real danger. The blood test results offered me a get-out, a chance to listen to my body's injuries with little loss of face. But I knew inside that there was another way. The risks were manageable if I had further tests with each subsequent race. Giving up is fine if you really have no choice. But if I stopped when I could have done better, I would regret it for the rest of my life.

During the flight, things began to improve. I felt less ill and was able to eat and drink a little. I was still peeing smoke and my legs remained pretty weak but with some sleep – thanks to our plush First Class seats – I was definitely more refreshed by the time we landed at Heathrow in the early hours of 31 October. Nevertheless, the run through London remained a bleak prospect.

We had always wanted our home leg to be special. And what could be better than following the route of the first ever marathon of the modern era? For the 1908 Olympics, runners had set off from Windsor Castle, so that Edward VII could watch the start, and finished at the sparkling new White City Stadium. We would do the same, although the fine old stadium had now become a BBC car park. The only drawback was that whereas the original runners passed through countryside over minor roads, we would run along major city arteries clogged by London's morning traffic. Still, the start was close to Heathrow and we would have time for a quick breakfast at an airport hotel before changing to go to Windsor. The hotel also gave me a chance to weigh myself again. I found that I now retained a mere four litres of extra fluid rather than the six of the day before. At least things were not getting worse.

We arrived at the castle walls at 7 a.m. Waiting for us were the by now expected TV cameras and reporters. More importantly there were also our friends, some of whom were planning to join us on the run itself. Leading our group was Hugh Jones, one of the best marathon runners that Britain has ever produced. He had gone to great lengths to track the original 1908 marathon course as closely as possible and he would serve as our guide. Also there to speed us along were Bruce Tulloch, another distance-running legend and one of the first to write books on the subject, and John Aggleton – one of my oldest friends and a man with whom I had run intermittently for more than thirty years.

Leaving the castle we crossed the Thames to Eton but all too soon we had left Ran's old school behind and moved into the less salubrious roads of west London, keeping to pavements or anxiously hugging the verge of fast urban highways. Then, after nearly an hour of running, which did not cause me too much trouble, we reached Uxbridge tube station where *Runner's World* magazine had arranged for another thirty or so runners to join us. It had been too dangerous for so many to run the route thus far but now they would stay with us all the way to the finish.

It was during the second hour of the London run that that dreadful strange fatigue hit me once more. It was nowhere near as bad as it had been in Singapore but I still knew that I would fail to run the whole way. I decided that I *must* complete at least half the marathon without stopping and so pushed on, despite my body's protests, until nearly two and a half hours had passed. Then, once I was sure that the half-marathon point was passed, I simply had to succumb. A small number of the runners with us, including my friend John, decided to stick with me. While Ran continued with his bid to run every step I was back to my running-and-walking routine. This did not slow us down too much, however, and unlike the day previously, I did not feel unwell. In the end I guess that I ran about three-quarters of the distance, completing the whole thing in five hours and fifteen minutes. Once again, I had been beaten easily by my partner who had come in just over half an hour ahead.

I had at least managed to run the last few hundred yards continually – and pretty swiftly for a man with damaged legs. My two kids, Callan and Tarn, had been waiting for me close to the end of the route and, with them alongside, my leg muscles – which five minutes before had disobeyed all commands – demonstrated that they were still capable of doing something. This was both pleasing and depressing. Perhaps, if motivation from my family could increase my performance, Ran's ability to keep going when I could not was a reflection of his greater mental strength. I took solace in those measures of muscle damage; they made concrete the physical limitations of my efforts.

The end of our London run was marked by the pleasure of meeting my wife Thea, my father and more friends, along with the less pleasurable blood test, this time arranged by the British Heart Foundation. But my good spirits did not last. Exhausted, I stumbled off for a warm shower in the nearby BBC headquarters but the dressing-room plumbing was faulty. The drenching was a chilly experience which added to a real mental low as I left my family for Heathrow. We would be starting the next leg of the challenge – Cairo – later that same day and the prospect filled me with horror. There was, however, one plus note: the results of our latest blood assessments showed Ran's CK figures were down to perhaps twenty times normal, with mine following in proportion at around two hundred times.

*

Although running our fifth and sixth marathons in such quick succession had always been the most daunting part of our timetable, as we approached Cairo at dusk I felt no worse than I had done on landing at Heathrow that morning. Indeed, in some ways I felt better and, at the time, I couldn't understand why. As we drove through the city, over the Nile and out to the pyramids, we were struck by the almost overwhelming noise and bustle. This is pretty much permanent for Cairo, especially during the feast of Ramadan when, for a whole month, daytime fasting is countered by an evening of feasting. We'd arrived right in the middle of the festival, and the whole city was heaving in one big party.

I had always wanted to see the pyramids and as I stepped from our vehicle they were suddenly before me. They met all expectations: huge, calm, simple forms, lit up against the backdrop of desert sky. Only the Sphinx was a mild disappointment, smaller than I had imagined since first reading of it in childhood. I would have loved to stay longer, at least had a quick tour, but there was no time. The brevity of our visit would make even the fastest sight-bagging tourist blanch.

At the foot of the pyramids, the Egyptian authorities had gone to town on our behalf. Mrs Mubarak, Egypt's First Lady, had asked us to raise money for her charity, Women for Peace. She had sent a personal spokesman to wish us well and her backing had brought the press corps out in force. There were also about sixty or so runners, some native Egyptians and some from the British expat community, waiting to accompany us. Not all of those would be successful – in fact, only about a quarter of those who set out were to finish, a fact I found quite encouraging. It is always good when runners much younger than you, and in this case trying only one-seventh of you challenge, cannot keep up. Certainly, after five marathons, it gave a much needed psycho-logical boost.

We ran through Cairo in an atmosphere of carnival and chaos. Up at the front Ran and I led the way, flanked by the other runners and a mix of ramshackle vehicles. There were mobile television crews, photographers and journalists in cars, some in the back of pickups, some on motorcycles. In addition, we were being followed by a bus complete with lavatory – a true Portaloo. Less promisingly, there were also two ambulances, one for each of us. At regular intervals, these would draw up next to us as we ran, the paramedics beckoning urgently for us to come and ride with them in order that we might rest. They seemed to be missing the point.

For reasons I still cannot understand, all of the vehicles with us tried to follow in line abreast. This ensured that the main highway to the city – three or even four lanes across – was completely blocked at all times. As a consequence a huge queue of hooting traffic built up behind; at one stage it apparently stretched back for more than a mile. The only transport that could make it past were the underpowered Egyptian mopeds, each of which carried at least two riders. They would draw alongside for a few hundred yards while their riders called across to us or apparently swapped abuse with the occupants of several police cars who were singularly failing to maintain order. Vehicles of plain-clothes security men also added to the havoc. On two occasions our supposed protectors even jumped out ahead of us and started fighting with the pedestrian well-wishers who lined the way. Overall, it was chaotic but filled with cheers, claps, hooting and warmth. The Egyptians took us to heart and we were thrilled by our reception. Cairo was a fantastic experience. In the end, of all the runs, it was by far the most fun.

I suppose it was inevitable: once again I failed to run every step of the marathon. Still, I was pleased to find things easier in Cairo than in London. In fact, I probably ran 80 per cent or more of the sixth leg, which showed in my time of four hours, forty-two minutes – not only faster than earlier in the day but also quicker than the Sydney course where I had run every step. I was also delighted to come in just twenty-five minutes after Ran, whose time of four hours seventeen was actually faster than any of his other marathons except the initial wind-assisted effort in Patagonia. Things were looking up. All we had to do now was to get to New York and cover the course in seven hours. Completion of our imaginative challenge was moving steadily towards reality.

*

I have puzzled over the variation in my fitness through the first six of these marathons, wondering why the muscle damage appeared to be maximal in Singapore and why, excluding the two earliest runs, I was at my best in Cairo. Three possibilities come to mind which, in reality, may have combined.

First, it is well known that the type of muscle action that causes most muscle damage is 'eccentric work' – the action of a muscle pulling while simultaneously lengthening rather than contracting. The best example of this comes from running downhill, when the quadriceps muscles on the front of the thigh have to brake your speed instead of propelling you forwards. They are therefore actively working while your leg bends instead of straightens. It is possible that the main damage to my muscles came from the hills in the Falklands, followed by the steps *down* from the Sydney Harbour Bridge. The leaking membranes then led to increasing swelling over a couple of days, explaining why things were at their worst in Singapore. Furthermore, since repeated downhill running offers some protection from this type of damage, this might explain why Ran was relatively spared. He lives on Exmoor and had great experience of downhill running from his local training.

The second possible explanation goes back to the topic of heat – I have already described how hyperthermia could compound muscle membrane damage causing an increase in swelling, leaks and dysfunction. This clearly applied to Singapore alone, yet the greatest degree of water retention came *before* the Singapore run. Furthermore, the myoglobin in my urine was at its most obvious on leaving Sydney. I also felt at my lowest on the flight between those two cities. It therefore seems unlikely that it was the heat alone that made Singapore such a misery.

Last, at least a part of my variable performance could well have been down to circadian rhythms. By the time we had completed the first six runs, we had chased the sun three-quarters of the way around the globe, creating about eighteen hours of jet lag. In general, our bodies can recover by only about one hour each day. It is therefore likely that on the day we ran in London and Cairo – the sixth day of our challenge – we were about twelve hours out of phase. This put us wildly out of step with our biology; as we passed through Eton, our bodies must have thought it was the middle of the night and all hormone support was at a low. Ironically, when we ran through the dark in Cairo we had awoken and our systems were perfectly prepared to help with our efforts.

*

New York, 2 November and the start of our seventh marathon. What a contrast to our first. Then, we had stood alone in early-evening sunshine, on a tiny track in Patagonia. Now, less than a week later, we were just two among tens of thousands waiting on the Verrazano

Narrows Bridge. Ahead, although I had never visited this huge city, we were to pass through places that I knew well from films and TV. Instead of unnamed beach, track and moor we would run the streets of the Bronx and the canyons of Manhattan to finish at last in Central Park.

It was marvellous to be among the crowds and I actually felt fit. For the first time since Sydney, we had now had more than twenty-four hours' rest. Better still, several of those had been on the ground. Although sponsored flights had forced us to fly by a long route from Cairo, via London to New York, we had still arrived at JFK airport the previous evening. There we had been met by Mike Kobold, a good friend of Ran's, who arranged a quick transfer to our downtown hotel. We settled in by 10 p.m., knowing that we had to be up again by five to reach the Staten Island start before the roads closed. Excited by the prospect of our final day and once again subject to our highly confused body clocks, we found that sleep did not come easily. Nevertheless, it was *so* good to be in a bed.

Among our 30,000 fellow runners, we had three there to help us. Steven Seaton was a long-standing friend of Ran's with whom he had shared several punishing running events in the past. He was also the editor of *Runner's World* and, as well as organising those runners who had joined us back in London, he had used his widespread contacts to become instrumental in much of our seven-marathons planning. Now he had flown over to run with us as our pacemaker, a role he had adopted frequently when helping groups of *Runner's World* readers complete the London Marathon. The second of those to accompany with us was Ran's nephew, Tony Brown, a US doctor who had also run marathons. I had not met him before but he seemed friendly, calm and supportive. I felt as if I could rely upon him. Finally, there was Mike Kobold. Although slim and almost unbelievably enthusiastic, I was less certain he could help. It was clear on talking to him that he was no runner and his training to join us for this event had been extremely limited. Perhaps his occupation helped, I thought. Was he on his feet and moving all day? No, he was a watchmaker – a bench-based craftsman, producing his own upmarket, handmade brand. It seemed unlikely that he would be with us for much of the distance.

Before the off, Steven asked how long I thought the run might take. I considered carefully before answering: I did not want him to push too hard. But I felt good and more rested, and believed I could run reasonably. Encouraged by the big-race atmosphere, I predicted four and a half hours but really expected faster. Why shouldn't we come in under four?

What a foolish concept. The first thing I had done when I had got out of bed that morning was to weigh myself. I was still four kilos over my starting weight. I should have realised that this meant my muscle fibres were still damaged. Worse, my feet now sported several blackened nails and a toe that had lost both nail and skin back in Singapore. It was a macerated mess and hurt dreadfully even when standing still. To think of this event as easy was self-delusion at its best.

On starting, my stupidity was brought home to me within moments. The gentle incline of the Verrazano Narrows Bridge felt like a steep ascent late in a mountain marathon and, although I could run, a brisk walker would not have been far behind. The echoes of the start gun had barely died before I dropped my four-hour dream, replacing it with a more modest aim. I would try to salvage a bit of pride by running every step.

For the first few miles, although I found it hard, running was easily possible. I even enjoyed the experience and felt well enough to look around at my surroundings and fellow competitors. The nature of the runners was somewhat different to my previous big-race experience. They had the air of men and women who took the whole thing very seriously. Unlike in the London race, there were few runners dressed as Mickey Mouse, rhinos, or lifeboats. Even a group of men running in camouflage military fatigues turned out to be members of the Parachute Regiment from Aldershot. Apparently the tradition of sponsored running for charity, with extra incentive for 'additions' that make life even more difficult, is very British. We did pass and exchange 'good lucks' with the paras but can claim little kudos for doing so. They were carrying fifty-pound backpacks.

The level of media attention we received was a considerable surprise. Before we set off, we had been interviewed by CNN and other New York TV stations, who were clearly making us one of the two main race stories. Our competition was none other than the rap star P. Diddy (formerly Puff Daddy) who was also running. For the crowds, this meant that any significant TV action must relate to his passing and there was a huge surge in interest as we approached. This led to some amusement en route. As Steven Seaton later put it, the crowd, waiting expectantly for the flamboyant rapper 'showed obvious disappointment to find only a pair of haggard old men'. I am sure he was right, although I found 'old' a touch harsh.

At ten miles I was struggling and by halfway I was wrecked. I was still just about running but knew I wouldn't be for much longer. My determination had been caught by reality and my limbs would simply not answer commands. Any incline, however gentle, was purgatory and with a little more than ten miles left I became very slow. Ran was still forging ahead but he soon hit on a generous if painful (for him) solution. Unlike previously, he had determined that we should cross the line together. Every few hundred yards he would turn round and drift back towards me. Other participants, spotting this figure moving in the wrong direction, must have questioned his sanity. Had they known that this was now his seventh marathon in seven days, he might have been committed.

Even with intermittent walking, I began to wonder if I could finish. My legs at times were literally buckling beneath me and this led to an intermittent sudden stagger. Worse, when I did so, I could not help but yelp with pain. Signals from macerated toes can be tolerated when fitted into a rhythm of expectation, but are harder to manage when attempting not to fall. The sudden movement mashes your toes into your shoes.

With increasing concern, Ran's nephew Tony stuck by my side urging me on and so did Mike Kobold. To my amazement, this unfit watchmaker was still with us, fetching me drinks, and telling surrounding runners of my story so that they too would offer support. Although by halfway he was also suffering, there was no way that he wasn't sticking with me. He therefore found for himself those survival reserves to draw upon and, just in case mine were finally run down, he offered me a real incentive to finish. If I could drag myself to the final line (and so by default drag him), he would give me one of his exclusive watches. He was a man of his word and a suggestion that I thought had been an empty heat-of-the-moment throwaway was fulfilled as a promise. A few weeks later, a beautiful Kobold chronometer arrived at my home by courier.

Central Park was a scene of wonder. Each side of the road was lined with tens of thousands of well-wishers and although we had ended up going slowly there were still thousands of other runners around us. For me, its gentle hills were still utmost tests of resolution, but now there was a difference. From the moment I entered the park gates, I knew for sure that it was all over. Even if I had to crawl, I would cover those final two miles in the two hours that were still left before our clock hit seven full days. I had not done as well as I had wished, for I had not run every single step. But at the outset, I did not really think that I would make it at all. Now I was about to complete the undertaking.

And then there was Ran. He had never stopped running, not even for a single step, in any one of the seven marathons. And, just as with the efforts of Helen Klein in the Utah Eco-Challenge or my own father in the British Columbia race, I had again been privileged to witness something special. As I saw him waiting, trotting on the spot a few hundred yards from the finish, I was so grateful. When we finally crossed that seventh line together, it was a moment to cherish. I cannot thank him enough.

In the end the New York run took us five hours and twenty-four minutes, longer than all the others except for the nightmare of Singapore. But the time did not matter. Surrounded by the press, TV and well-wishers, Ran and I shook hands and hugged to celebrate our success. He too had never really thought that we could bring off our audacious plan. Now we had not only done it, we had even enjoyed some of it, and more importantly we had raised money for a host of good causes. As we walked out of Central Park, I borrowed a mobile to phone home but was suddenly choked with emotion. My wife answered and for several moments heard nothing but tears.

*

We left New York that evening for the last stage of our 45,000-mile journey. On board the plane, I could not sleep and found myself looking back to the start of our project. Even now it seemed surprising that we had achieved so much. With thought the explanation became simple. Ran and I had both known through experience that if we trained we could acquire those adaptations of endurance that allow men and women to run huge distances easily – the strong heart, the good lungs and the enlarged slow muscle fibres that can just go on and on. We also knew that when our endurance flagged, we could call on the reserve that has been granted to all for use in desperation. Once recognised, this makes the human the ultimate endurance machine. It explains why someone like myself, with very ordinary athletic talents, can do so well.

In the end it had turned out to be a question of mind over matter. Making our challenge happen demanded an unusual attitude from the moment of its inception, through months of logistics, setbacks and training, right up to the end in New York. When, following our success, many experts in both the USA and Britain expressed disbelief at what we had achieved, they did not realise that they could have done it too. The difference is only one of perception. Whereas most people look at very big challenges, whatever the field or their walk of life, and start from the position 'I can't', Ran and I make a simple word substitution and say '*Why can't*'. 'I can't run seven marathons' easily transforms into the question 'Why can't I run seven marathons?' Once it was asked, we felt obliged to find the answer.

Sources and Further Reading

THESE SUGGESTIONS are for those interested in exploring further the subjects raised in this book. It is neither a full bibliography nor a complete list of scientific references, although the list does provide useful starting points for those wishing to examine the primary literature. They are listed with comments where appropriate. Scientific references are given in standard format using recognised abbreviations for the journal titles. Publishers are in the U.K. unless otherwise indicated.

Human Evolution

Richard Dawkins' *The Selfish Gene* (Oxford University Press, 1982) and *The Blind Watchmaker* (Longman, 1986) give comprehensive accounts of the workings of natural selection and evolution, particularly emphasising how genes and the DNA dictate everything that happens in the evolutionary process. This approach has been criticised recently by Stephen Rose in *Lifelines* (Penguin Press, 1997) as being too reductionist. Rose suggests that the forces of natural selection operate at a much higher level than the single genes, selecting for the whole organism within the context of its environment.

Jared Diamond's *The Rise and Fall of the Third Chimpanzee* (Radius, 1991) sets out a thorough and very readable account of human evolution, pre-history and the development of civilisation. It also describes in much more detail than I have done the relationship between man and the chimpanzee, giving details of how the closeness of our relationship has been demonstrated. *The Naked Ape* by Desmond Morris (Jonathan Cape, 1967; illustrated edition, 1989) is now a classic, providing a summary for laymen of human development. It discusses the reasons for our becoming upright, as well as the development of human intelligence, tool use and speech. It is, however, heavily criticised by Elaine Morgan for being too male orientated in *The Descent of Woman* (Souvenir Press, 1972). Her book gives a most readable, if rather biased, presentation of the 'marine theory' of human evolution – the idea that we spent a significant period after dividing from the chimpanzees as a predominantly water-dwelling mammal. A more balanced analysis of the pros and cons of this theory is reported in the proceedings of a conference on the subject: *The Aquatic Ape: Fact or Fiction* (eds M. Roede, J. Wind, J. Patrick & V. Reynolds, Souvenir Press, 1991).

Sports Physiology and Nutrition

Two recommended books giving broad coverage of the subject of muscle function and training are: *Textbook of Work Physiology – Physiological Bases of Exercise* (P.-O. Astrand & K. Rodahl, McGraw-Hill, New York, 1977) and *Physiology of Sport and Exercise* (J. Wilmore & D. Costill, Human Kinetics, Champaign, Illinois, 1994). An excellent and reasonably up-to-date review of the interaction between nutrition and exercise is given in *Foods, Nutrition and Sports Performance* (eds C. Williams & J. Devlin, Spon, 1992) and I have recently written a review of the dietary needs for very protracted physical effort – *Nutrition for prolonged endurance exercise* (M. Stroud, Proceedings of the Nutrition Society, 1998).

Heat and Cold

By far the most complete coverage of the human physiological responses, adaptation and acclimatisation to heat and cold can be found in the book *Human Performance Physiology and Environmental Medicine at Terrestrial Extremes* (eds K. Pandolf, M. Sawka & R. Gonzalez, Benchmark, Indianopolis, 1988). A much briefer review is outlined in a chapter written by myself and entitled 'Environmental Temperature and Physiological Function' in the book *Seasonality and Human Ecology* (eds S. Ulijaszek & S. Strickland, Cambridge University Press, 1993).

Polar Expeditions

Accounts of various Polar expeditions to which I refer in Chapters 4 and 6 include: *In the Footsteps of Scott* (R. Swan & R. Mear, Jonathan Cape, London, 1987); *Scott's Last Expedition* (Capt R. Scott, Smith, Elder, London, 1913); and *The Worst Journey in the World* (A. Cherry-Garrard, Chatto and Windus, 1922).

The full story of my own expedition across Antarctica with Sir Ranulph Fiennes is given in *Shadows on the Wasteland* (M. Stroud, Jonathan Cape, 1993). Accounts of journeys across the sea-ice to the North Pole are given in Robert Swan's *Icewalk* (Jonathan Cape, 1990) and Ranulph Fiennes' *To the Ends of the Earth* (Hodder, 1983).

Scientific papers covering the physiological studies made on my own Polar journeys include: 'Nutrition and energy balance on the "Footsteps of Scott" expedition 1984–86' (M. Stroud, Human Nutr. Appl. Nutr. 41A: pp. 426–433, 1987); 'Measurements of energy expenditure using isotope-labelled water ($^2H_2^{18}O$) during an Arctic expedition' (M. Stroud, W. Coward & M. Sawyer, Eur. J. Appl. Physiol. 67: pp. 375–379, 1993); 'Nutrition across Antarctica' (M. Stroud, British Nutr. Found. Bulletin. 19: pp. 149–155, 1994); 'Protein turnover rates of two human subjects during an unassisted crossing of Antarctica' (M. Stroud, A. Jackson & J. Waterlow, Br. J. Nutr. 76: pp. 165-174, 1996); 'Thermoregulation, exercise, and nutrition in the cold' (M. Stroud) in *Physiology, Stress and Malnutrition: Functional Correlates, Nutritional Intervention* (eds: J. Kinney & H Tucker, Lipincott-Raven, New York, 1997); 'Energy expenditure using isotope-labelled water ($^2H_2^{18}O$), exercise performance, skeletal muscle enzyme activities, and plasma biochemical parameters during 95 days of endurance exercise with inadequate energy intake' (M. Stroud, P. Ritz, W. Coward, M. Sawyer, D. Constantin-Teodosiu, N. Brown, P. Greenhaff & I. Macdonald, Eur. J. Appl. Physiol. 76: pp. 243-252, 1997).

Nutrition, Exercise and Illness

Excellent sources of information on the benefits of exercise to all aspects of health including heart disease, obesity and ageing are *The Benefits of Exercise: the Evidence* (P. Fentem, J. Turnbull & J Bassey, Manchester University Press, 1993) and *Physical Activity, Fitness and Health* (C. Bouchard, R. Shepard & J. Stephens, Toronto, 1994).

The literature discussing more specific details of factors underlying heart disease is enormous. *Preventing Coronary Heart Disease in Primary Care: the Way Forward* (National Heart Forum, H.M.S.O., 1995) would be a good starting source and a review of the benefits of physical activity on heart disease is given in 'Exercise in the Prevention of Coronary Heart Disease: Today's Best Buy in Public Health' (J. Morris, Med. Sci. Sports Exerc, 26: pp. 807-814, 1994). The British Nutrition Foundation have published numerous books and articles on the interaction of diet and coronary problems which include: *Diet and Heart Disease – A Round Table of Factors* (ed M. Ashwell, BNF, 1993); *Fish and coronary heart disease – has the tide turned?* (M. Sadler, BNF Bulletin 20, 1995); *Homing in on homocysteine* (U. Arens, BNF Bulletin, 20, 1995). The *Allied Dunbar National Fitness Survey* (Sports Council and Health Education Authority, London, 1992) provides figures for the increasingly sedentary nature of the UK population, while further figures from Britain and beyond are cited in some of the works on the causes of obesity I have listed below.

Why We Get Sick (R. Nesse & G. Williams, Times Books, New York, 1994), published in the UK as *Evolution and Healing. The New Science of Darwinian Medicine* (Weidenfield & Nicholson, 1995), examines many aspects of human health and disease from an evolutionary point of view, including brief discussion of heart problems and obesity. However, the authors pay surprisingly little attention to the role of declining activity levels in any of their discussion. Thorough general histories of the changing face of disease in society can be found in Roy Porter's *The Greatest Benefit to Mankind: A Medical History of Humanity from Antiquity to the Present* (HarperCollins, 1997) and *Diseases, their Emergence and Prevention* (eds. Trowel and Burkitt, Harvard University Press, 1981.).

The Causes of Obesity

In *Energy Balance and Obesity in Man* (Elsevier and North Holland Press, 1974) John Garrow gives a complete description of the interaction of dietary intake, resting metabolism and activity levels underlying the development of obesity. Once again, articles and reports published by the British Nutrition Foundation provide many useful starting points to further reading: *What should public health policy be towards 'overweight'?* (B. Tuxworth, BNF Bulletin 19, 1994); *Body Weight and Health* (ed M. Sadler, BNF, 1996); *Fat and obesity* (J. Westrate, BNF Bulletin 21, 1996). Two excellent articles emphasising the fact that the obese do not have slow metabolism and hence abnormally small dietary needs are 'Don't Blame the Metabolism' (A. Prentice. MRC News, Autumn 1995) and 'Obesity in Britain: Gluttony or Sloth?' (A. Prentice & S. Jebb, Br. Med. J. pp. 311, 1995). A specific review of the recently postulated role of leptin in obesity can be found in *Leptin – the 'new' player in energy balance and obesity* (P. Trayhurn. BNF Bulletin 22, 1997).

Exercise and Ageing

Many of the books listed in the second and fifth section above contain much information on the benefits of exercise with age. More specific information can be found in scientific papers including: 'Endurance training in older men and women. 1. Cardiovascular responses to exercise' (D. Seals, J. Hagberg, B. Hurley, A. Ehsani & J. Holloszy, J. Appl. Physiol. 57: pp. 1024–1029, 1984); 'Effect of age and training on aerobic capacity and body composition of master athletes' (M. Pollock, C. Foster, D. Knapp, J. Rod & D. Schmidt, J. Appl. Physiol., 62(2): pp. 725–731, 1987); 'Fifteen-year changes in exercise, aerobic power, abdominal fat, and serum lipids in runners and controls' (B. Marti, M. Knobloch, W. Riesen & H. Howald, Med. Sci. Sports Exerc. 23(1): pp. 115–122, 1991).

Acknowledgements

MANY PEOPLE have contributed indirectly to this book – educating me in the medicine and science underlying it during my time in the Anthropology Department at University College London; the Medical School at St George's Hospital, London; the Centre for Human Sciences, Farnborough; and in my current post at the Institute of Human Nutrition, Southampton. Particularly important in this respect were Prof. Mike Stock, who first pointed me towards research in human physiology and metabolism, and Prof. Alan Jackson, who, in his teachings on nutrition, has made me think in terms of teleology. Thanks are also due to all those who made my expeditions and other outdoor activities possible: my father, who started it all off by taking me to the hills as a child and teenager; Robert Swan and Roger Mear, who set up the 'Footsteps of Scott' and so involved me in major expeditions; Ranulph Fiennes, who was both an inspiration and a wonderful companion on our many trips together; Lawrence and Morag Howell, who supported those trips I made with Ran; Chris Lawrence, who introduced me to the Sahara race and hence to ultra-distance events; and Mary Gadams who first got me involved with the Eco-Challenge.

Lastly, I owe a huge debt to those who were directly involved with the writing of this work: my wife, Thea, who helped me put my thoughts together by discussing things and reading my draft efforts, and who coped with so much while I was busy, and my children – Callan and Tarn – who had 'daddy behind the computer' for weekend after weekend. Tony Colwell, my editor at Jonathan Cape, was patient, critical, methodical and encouraging in turn, and I am sure that without him the book would never have been completed.